EXHIBITION
Poultry Keeping
DAVID SCRIVENER

THE CROWOOD PRESS

First published in 2005 by
The Crowood Press Ltd
Ramsbury, Marlborough
Wiltshire SN8 2HR

www.crowood.com

British Library Cataloguing-in-Publication Data
A catalogue record for this book is available from the British Library.

ISBN 1 86126 739 8

Acknowledgements
Most of the photographs were taken by the late John Tarren. Thanks are due
to his wife, Mrs Joyce Tarren, for allowing me to buy his collection. I must also
thank Mr Andrew Sheppy for helping with the appendix covering poultry
organizations; and Mr Steve Fisher of Fisher's Woodcraft, Church Farm,
Church Lane, Fishlake, Doncaster DN7 5JP (tel: 01302 841 122, fax: 01302
840 088, email: smpltd@aol.com) for the photographs of the poultry houses.

Typefaces used: Goudy (*text*),
Cheltenham (*headings*).

Typeset and designed by
D & N Publishing
Hungerford, Berkshire.

Printed and bound in Great Britain by CPI Bath.

Contents

Introduction

Most beginners to domestic poultry keeping start with home egg production as their main motive. For this purpose they keep either commercial hybrids or those pure breeds noted for attractive eggshell colours, usually Araucanas, Marans or Welsummers, possibly Crested Legbars, Welbars or Barnevelders. In future years the Penedesenca from Spain might be added to the list of popular beginners' choices as they, too, lay very dark brown eggs.

Some people are content to stay with this type of flock, and if they are not permitted to keep crowing cockerels, they have little choice. Once involved with poultry, books and magazines are studied, and farm parks and poultry shows are visited, and so the interest in the amazing range of shapes, sizes and colours of domestic breeds of poultry and other birds develops. Many home menageries grow as more stock is added – and sometimes rather too much if enthusiasm goes too far! At this stage some people begin to realize how rare some breeds are. They might read about a certain variety, decide to buy some, and then discover that there are very few specimens in the country, and none of these very near. It is to be hoped that most of these people will quickly realize that these breeds need more breeders to conserve them, and so will make the effort to do so.

Exhibition Poultry Keeping is primarily intended for the minority who progress to the next stage: that of serious breeders and exhibitors. It will, however, be just as useful for the many who prefer to stay with their menagerie. Breeds with heavily feathered feet, head crests and very long tails need special care if they are to look their best, and these techniques are not usually included in *How to Keep Hens*-type books. Even if it is not intended to show them, these breeds will surely be appreciated better if the features for which they were bought are in good condition.

Some poultry varieties have intricate plumage patterns that can degenerate if the best birds are not selected for breeding each season. This can be a complex process with those varieties needing to be 'double mated', a term that has caused a great deal of confusion among novice poultry keepers. This is fully described in this book, as are many other details that everyone intending to maintain a flock of a pure breed needs to know. As with the feather-footed and other types mentioned above, correct breeding methods are needed to keep pure breeds properly, whether they are shown or not. Farm parks rely on sales of stock for a significant part of their income, and only those supplying good quality birds and hatching eggs will have people coming back for more.

It does not need an expert to realize that show poultry, especially white plumaged breeds, have been bathed, blow-dried and manicured; they were not just picked up from a muddy run looking like that. Show preparation is therefore fully described here. Although exhibitors are happy to spend time giving their birds this beauty treatment, very few of them expect to win major awards, such as 'Best in Show', regularly. They do take the judging of the breed classes very seriously, however, and can get

The Indian Game Club stand at the PCGB National Show. Many of the breed clubs have similar stands at the National (Stoneleigh) and Federation (Stafford) shows, plus a few at other events, Reading Bantam Club Show, for example. They are a useful meeting place for existing members, and to attract new ones. Novice pure-breed poultry keepers might decide to keep a different breed if there is no club stand for the variety they originally had in mind.

very upset if the judge clearly does not understand the breed, or at least agree with the owner's interpretation of the standards. There is seldom much, if any, prize money to be won nowadays, but fanciers still think it important that the right birds win. It is usually the case that direct competitors with the same variety are also the closest friends, and during judging, groups of breeders will be seen discussing the finer points of their feathered favourites.

Compared to other hobbies and leisure activities, poultry showing has a very limited following; indeed, a lot of people have never heard of it, 'You show *chickens*?', being a common response. Nevertheless, it is perhaps rather more popular and widespread than many people imagine, and shows are held all over Europe,

the USA, Canada, Australia, New Zealand, Japan, South Africa, and probably a few more countries besides. As there are poultry keepers all over the world, there is clearly scope for the hobby to expand. It is known that there are a few pure-breed enthusiasts scattered all over individual countries, in several South American countries for example, who might one day organize themselves into show societies.

It is also a very socially inclusive hobby, with fanciers of all ages and backgrounds. You do not have to be particularly fit to handle chickens or to clean out hen houses, so fanciers can range from nine years old to ninety, including those with some physical limitation. Also it isn't necessary to be wealthy as many champion birds have come from garden flocks.

CHAPTER ONE

A Brief History
of Poultry Shows

The first significant poultry show in the world was held in the grounds of London Zoo in 1845. More details of this event are given below, but first the forerunners of this event must be mentioned. Private competitions had been held for many years before 1845 for breeders of Sebright Bantams, White-Faced Spanish and Hamburghs, the latter then having several local names. These early competitions were usually held in pubs.

The birds that were later standardized as Derbyshire Redcaps, Black Hamburghs, Gold-Spangled Hamburghs, Silver-Spangled Hamburghs and Old English Pheasant Fowl were the subject of pub competitions in their local areas. The eventual Black Hamburghs were then called Black Pheasant Fowls, and some of the Spangled birds were named Lancashire Moonies and Yorkshire Pheasant Fowl.

These were fairly informal events, so there are not many detailed records in the poultry books. They may have been reported in local newspapers at the time, but this has not been researched. It is believed that assistants held the birds for the judges to examine, although it is possible that these clubs may have had a few judging cages, and judged the birds a few at a time. Most of the competitors were normal working people, which is confirmed by the very practical prizes offered: copper kettles and other domestic utensils, or perhaps a large joint of meat.

The ideals used at these events must be the first-ever breed standards. Wingfield and Johnson included them in their 1853 *Poultry Book*. The standard for Blacks is reproduced here to give an impression of the events; the standard for Spangleds was rather longer.

For Black Hamburghs/Black Pheasant Fowl:

Points	Marks of feathers, etc., considered best
1st Comb	Best double, best square, most erect, and best piked behind.
2nd Ears	Largest and purest white.
3rd Colour	The best and richest glossed green-black.
4th Legs	Best and clearest blue.
5th General appearance	Best feathered hen.

Very similar gatherings of White-Faced Spanish breeders were held much further south, London and Bristol being the two main centres. This breed needed, and still needs, protection from excessive sunshine, biting winds and frosts. These fanciers were pioneers of the arts of show preparation and spent many hours pampering their pets to ensure they would appear with the purest white, smoothest possible faces. Most were kept in suburban gardens or tiny backyards rather than on farms.

Black Hamburgh, large female.

White-Faced Black Spanish, large male.

Despite their name, the Spanish breed is believed to have been developed in what is now Belgium and The Netherlands, areas that were under Spanish rule for much of the fifteenth and sixteenth centuries. The White-Faced Spanish would have been brought over by refugees from the wars and religious persecutions of the seventeenth century. If these original breeders held any competitions, the history of poultry-showing may extend back over four centuries.

Sir John Sebright (baronet) started to establish his famous breed of laced bantams in about 1800. He was Member of Parliament for Hertfordshire, and the family estate was at Markyate, a short distance south of Dunstable, near the Hertfordshire/Bedfordshire county boundary. At some time around 1810 Sir John had made enough progress, and attracted sufficient interest from other breeders, to form The Sebright Bantam Club. This was originally a private and very exclusive group. The annual subscription was two guineas for each colour, amounting to a prohibitive – for all but the very wealthy – four guineas for those who kept both Golds and Silvers. Most of the subscription money was given back as prizes at the annual show. The show was held on the first Tuesday in February, initially somewhere in Brick Lane, Spitalfields, London, later moving to the Gray's Inn Coffee House, Holburn. New members had to be proposed before they were admitted to the club.

It is not entirely clear if the present (2004) Sebright Club is a direct continuation of Sir John's original group: if it is, it is easily the oldest poultry breed club. W.F. Entwisle said that the original club was still going when he wrote his book *Bantams* in about 1890. By 1910 the Sebright Club subscription was still high, but at 10s 6d was affordable to at least the respectable middle classes; in 1926 the club was charging a much more democratic 5s. The pedigree of the current Sebright Club is possibly confirmed by the fact that the president in 1932 was Sir Guy Sebright. Sir John had died in 1846, his son was Sir Thomas, and there was probably at least one more generation before Sir Guy.

The ideal Sebright Bantam has remained the same for two centuries: clear lacing, rose comb, purplish face, short back and strutting carriage similar to that of a Fantail pigeon. The maximum weights allowed were 22oz (624g) for males and 18oz (510g) for females. We shall never know how close the early breeders got to this ideal as very few photographs survive from before 1900. Furthermore, the drawings and paintings are no guide, because all the artists depicted the ideal, and not the reality. One expert, J.F. Entwisle, wrote in 1926 that over the previous forty-five years there had been a great improvement in the lacing, but a loss of the short-backed, strutting carriage.

The 1845 London Zoo Show was different from the previous competitions in that it provided classes for a wide range of breeds:

Silver Sebright, bantam female.

- White, Speckled or Grey Dorkings
- Surrey Fowls
- Old Sussex or Kent Fowls
- Gold or Silver Hamburghs
- Spanish
- Malays and other Asiatics
- Gold- or Silver-Spangled, Black or White Polish
- any other variety
- Bantams, Gold- or Silver-Spangled (= Sebrights)
- Bantams, black or white (= Rosecombs)

All exhibits had to be a pen of a cock and three hens, and exhibitors had to provide their own cages. This event was successful enough for a second London Zoo Show (they were held outside, in the grounds) in 1847, and for a poultry section to be added to the 1848 Birmingham Cattle Show.

At these early events there were several distinct groups of exhibitors:

- Established fancy breeders of Hamburghs, Sebrights and Spanish, to which might also be included those of Polish and Rosecombs.
- Specialist table-bird producers of Dorkings, Kent, Surrey and Sussex fowls.
- Recent importers of exotic Malays, Cochins, Shanghais and others from the Far East.

Cockfighting was banned in Britain in 1849, resulting in the addition of game-fowl breeders to the exhibition scene. Although cockfighting continued in secret, the breeders needed a legal reason for continuing with their favourites.

THE FIRST AMERICAN POULTRY SHOW

Interest in breeds of poultry was increasing in America as well as in Britain, and this was stimulated by the arrival of the large, feathery Brahmas, Cochins and Shanghais from China: such magnificent chickens had never been seen before. It was only in this period that trade with China had started, a result of the Opium Wars. Among the new poultry keepers

was Queen Victoria, who had a large poultry house built at Windsor. Where the Queen led, a host of duchesses, countesses and Right Honourable ladies followed, and previously neglected hens became 'prize poultry', and started to change hands at extraordinary prices. This settled down a little after about 1855, the main period of chicken mania being permanently known as 'the Hen Fever'. Despite being proud citizens of the Republic, Americans seem to have followed the British Royal Family as much then as they do now.

On 8 October 1849, Dr John C. Bennett of Plymouth, Massachusetts, wrote to the editor of the *Boston Cultivator* to say that he intended to display the best of his collection of poultry breeds at Quincy Market, Boston, on Thursday 15 November. The editor, Mr James Pedder, decided to take this idea further, and turn it into a general poultry show. Quincy Market was not adequate for the big event he had in mind, however, so Boston Public Gardens were booked, and marquees hastily erected, the largest being about 22,500sq ft (2,100sq m). He acted very quickly, and the show was announced on 27 October, to be held, now for two days, on 15 and 16 November. To cover expenses, gentlemen visitors were to be charged 4 cents, while ladies and children went in free. His gamble paid off, as 10,000 spectators turned up, paying a total of $314.

The poultry breeders of the area provided an excellent display for the public to see, namely 1,423 pens of birds from 219 exhibitors. It is worth listing the breeds entered to see which varieties we can confirm have existed for over 150 years; they are:

Chinese, Cochin-China, Red Shanghai, White Shanghai (all recently imported from China), Bucks County (early American breed, roughly similar to Rhode Island Reds), Jersey Blues (early American breed, roughly similar to Blue Jersey Giants), Javas (early American breed, roughly similar to Black Plymouth Rocks), Malays, Dorkings, Spanish, Italians (Leghorns), Guelderlands (Bredas), game fowls, Creole Fowls (Dominiques?), Bolton

Greys (Hamburghs), Frizzled Fowls, Spangled Hamburghs, Pencilled Dutch fowls (as Pencilled Hamburghs? Friesians?), Crested Polands (Black, White, Gold-Spangled and Silver-Spangled), bantams (breeds not recorded) and recent crosses (including Dr Bennett's first Plymouth Rocks). There was also a selection of pheasants, turkeys, peafowl, guinea fowl, ducks, geese and swans.

Among those attending was US Senator Daniel Webster, mainly remembered in history for settling the US/Canada border with Britain when he was US Secretary of State.

BRITISH SHOWS DEVELOP

From about 1870, the main British show of each year was held at the Crystal Palace, until it burnt down in 1936. This show was the great annual meeting place of all the small livestock hobbies. In the early 1920s there were usually about 12,000 entries in total, of which about 4,500 to 5,000 were poultry, the other major sections being fancy pigeons and rabbits. After Crystal Palace burnt down, the event moved to Olympia. There were no shows during World War II, and showing livestock was much less popular in the nineteen-fifties; commercial poultry farming was developing into the industry it is today, with hybrid layers and broilers replacing the old pure breeds. The pigeon and rabbit breeders moved out to organize their own events and Olympia gradually became a trade fair for the new poultry agri-business; the pure breed poultry section remained, but was considerably diminished – the 1962 Olympia Show had just 2,194 live bird entries. There have been several years when fowl pest outbreaks caused all shows to be cancelled, but 2,194 must remain as a low point for the main UK poultry show.

Poultry were first included in the Dairy Show in 1876, when there were 473 entries in twenty-four classes. This rose rapidly to 1,100 entries in sixty-four classes by 1880, and then stabilized. The 1890 Dairy had 1,274 birds, despite a rise to 122 classes (and therefore more prize money) to enter them in. None of these figures is particularly impressive, but the event did gradually rise to become one of the 'classics' of every season. The 1923 Dairy

'What not to do at a show': the over-enthusiastic public as shown in How to Win Prizes with Poultry *by W. Powell-Owen, circa 1910.*

Show had a (then) record entry of 4,685 birds. No records are available to check if this was ever bettered, but the Depression, World War II and social changes post-war make this unlikely.

The third 'classic' British show each season used to be held at Birmingham's Bingley Hall every year from 1848. In the 1970s it moved to Stafford Agricultural Showground, and the 1921 show had an impressive 3,974 entries. After all the lean years, the Stafford Shows of 2000 to 2003 have attracted some 5,500 poultry, with separate pigeon, cage bird and fur (rabbits and so on) sections as well.

British poultry exhibitors have two really large shows to attend each year: Stafford is one, and the Poultry Club National Show at the Royal Showground, Stoneleigh, is the other. This also has about 5,500 poultry entries, but does not have space for anything else.

THE FANCY WORLDWIDE

Shows continued in the USA after the first Boston event, also up to the present day, although the sheer size of the country has been a problem for successful 'Nationals'. Entries of 12,000 birds have been achieved, however – and if there are enough people willing to travel the huge distances involved to these, there must surely be a lot more who are interested in poultry, but not able to travel. Shows are also held in Canada, but the travelling problems are even worse.

Fanciers in Australia and New Zealand hold some excellent shows, but have to contend with strict local authority regulations in many districts. They are also forbidden to import fresh stock from anywhere else in the world.

Several Japanese breeds have attracted a lot of interest all over the world, but poultry fanciers in Japan do not seem to be interested in breeds from around the world. In one respect this is good in that it provides a secure base for the continued existence of these Japanese breeds; but on the other hand, perhaps the Japanese might benefit from a broader view.

The real centre of poultry showing is Germany and The Netherlands. These two countries set the standards for how good a poultry show can be for the clubs in the rest of the European Union.

CHAPTER TWO

Choosing Your Breeds

Most readers of this book are probably happily keeping their chosen varieties, and have no intention of changing. This is as it should be, and the author has no particular intentions of persuading anyone to change to something else. Having said that, it is always pleasing when a new flock of a very rare variety is established. There are generally several reasons why people choose a particular breed: for example, the one grandfather kept, or perhaps the breed associated with the area. The more famous names in the poultry fancy, past and present,

have usually been associated with one breed; some of them are known to have kept their favourite varieties from youth until old age. Les Miles with North Holland Blues and John Croome with Campines and Autosexing Breeds were two such lifetime breeders, still remembered by poultry breed conservationists.

Not everyone is single-minded enough to stick to one or two breeds for decades, and even specialists often have a range of secondary breeds over the years, 'for a change'. Furthermore, having experience of a number of

North Holland Blue,
large female.

Silver Campine, large male.

very different breeds is extremely useful for show judges, helping them to fully appreciate the difficulties in exhibiting various types.

Some readers may be fairly new to the hobby, and although they may have several varieties already, have not yet made any permanent decisions. This is also as it should be, because the first few years in the fancy can be regarded as the 'learning phase'. Thus it is not unusual for some breeds to seem much less interesting after a year or so, and others that did not appeal when you were a beginner, become more fascinating as you gain experience. Marans are a popular beginner's choice, their dark brown eggs being the attraction, but fuzzy cuckoo barring is not a particularly attractive pattern. Conversely, some of the breeds that obviously require a lot of expertise to breed and prepare for showing may be a daunting prospect for beginners, but later on these same novices may decide to try a more 'difficult' breed.

Avoiding an unsuitable breed is probably more important than choosing exactly the right one. The chapters on housing, breeding and show preparation should be studied to ensure you can provide suitable conditions for whichever types you have in mind. Thus freedom to roam about a farm or smallholding, which will inevitably be muddy at times, is fine for some breeds, but not all – and the very active types, those that would love the farm, may be a disaster in a suburban garden because they will fly over fences and the cockerels may be frequent and noisy crowers.

The remainder of this chapter suggests some factors that should be considered when choosing which varieties to keep.

THE SELECTION PROCESS: QUESTIONS TO CONSIDER

Large Fowl, or Bantams, or Both?

If poultry-keeping space is very limited, bantams may be the obvious choice; though having said that, some of the smaller and very ornamental large breeds might be fine as well. Polands, Silkies and Sultans all need to be kept clean and dry, and a tidy group of sheds and covered aviary-type runs in a suburban garden would be as perfect for these as they would for a long list of ornamental bantam breeds.

Large fowl and bantam eggs can be incubated together in an incubator, and the resulting mixed group of chicks will be fine for the first few weeks. However, as they get older the large fowl youngsters will probably be too big and boisterous for small bantams to cope with, and separate rearing houses will be necessary; this will also be the case for groups of different ages. Once they all reach maturity, they can be swapped around into their respective varieties.

Where a much larger number of birds can be kept, it should be remembered that hundreds of tiny bantam eggs may be more difficult to sell than normal-sized hen eggs.

How Many Birds Should I Keep in Total?

Allowing one's enthusiasm to run too far, resulting in many more birds than the family can cope with, is a common reason for people to give up poultry altogether. This should not be a collecting hobby, something that is sometimes forgotten by breed specialists who decide to go for the whole set – every colour, large and bantam, of their chosen breed. Domestic arguments about the money, time and space now occupied by chickens will do no one any good.

So what is normal? This question may be of pressing interest to the less-than-enthusiastic family members of new poultry keepers. First, be aware that there is an annual cycle. The flock should be at its smallest in February, when there should be only the best breeding and show stock present. Chicks are hatched and reared through spring and summer, resulting in the highest numbers in August and September. As the young mature, they can be assessed. Many of the cockerels will have to be killed, the best of both sexes retained, and some sold, often in trios of one cockerel and two pullets. The number of birds present in August is typically about double the February population. (*See* Chapter 4, Housing and Equipment, for full details of separate summer/winter and adult/rearing options and systems.) Taking the February flock sizes,

anything under fifty birds is considered quite modest, fifty to 150 is probably normal, and much over that is 'big'. This is a hobby, after all, we are not poultry farmers. Anyone considering a large collection should check current local rules on the sale of eggs. The rules requiring egg grading and testing will probably continue to exempt small hobbyists' flocks, but the definition in terms of number of birds may change over time, and be different for each country.

How Many Kinds Should I Keep?

Equally one might ask, 'How many of each kind should I keep?' Given that every fancier has a limit to the total number of birds they can keep, these two questions are really two ways of asking the same thing. It is normal to start by buying a trio or two, and breeding more to gradually build up the flock. As explained in Chapter 13, Selection and Breeding Systems, a larger-sized flock gives more scope for selective breeding and avoidance of the harmful effects of inbreeding in successful exhibition strains. Again looking at the lowest February flock size, five males and fifteen females (making a total of twenty birds) should be taken as a minimum, and certainly those with very rare varieties, or with a very successful championship-winning strain, may choose to have a lot more. Moreover, if additional stock will be difficult to buy, the breeder will wish to be as self-contained as possible. In the case of a champion strain, there may be no shortage of that breed available, but none of the desired quality.

Thus a modest-scale fancier with space for about fifty adult birds should plan to keep two varieties, whereas larger-scale fanciers have more choice. For instance, there are several breeds in which it is normal practice to cross the colours, to some extent anyway: if one of these is chosen, a flock of, say, thirty birds might include two or three colour varieties.

How Many Chicks Need to be Hatched Each Year?

The more chicks the better in general, but there are certain factors to consider. For instance, inbred strains produce more uniform young-sters, so fewer chicks will be needed than with matings of unrelated birds. Some plumage pat-terns and comb types are notoriously unpre-dictable; Sicilian Buttercups are very difficult in both respects, which may explain why so few people keep them. A sound understanding of poultry genetics is necessary when programmes are undertaken to make new colour varieties of breeds; some generations, it can be predicted, will be fairly uniform, whilst others will be very variable, with only a small proportion likely to have the desired combinations of genes.

Whichever breeds or strains are involved, there will always be more cockerels and fewer pullets (females) than we would like. And before the youngsters can even be sexed, remember that not all the eggs will be fertile, and not all of the fertile eggs will hatch – so by the time you have realized that there are not many pullets in your flock, it will be too late in the year to start again, with large fowl anyway. It can therefore easily need over 100 eggs to be incubated in order to get ten good, showable pullets. Also more than ten pullets should be reared, because not all of them will be top quality. Some fancy breeds are notorious for very poor egg produc-tion, fertility and viability; Indian Game, Sebrights and Yamato Gunkei are probably the worst, so incubate every egg.

Some breeds can be successfully shown for several years, indeed they do not look their best until they are two years of age. Other breeds may have a normal lifespan, but only a short show career; light breeds with white ear lobes, especially Minorcas, are in this catego-ry, and a constant stream of young Minorcas are needed by regular exhibitors.

Do Some Breeds Win More Major Awards than Others?

Yes, they do. This does not mean that people are likely to keep a breed they do not like very much just to win a few cups; the flock has to be cared for every day, with even the keenest exhibitors only spending a few days each year at shows. For example, if a fancier likes a large white, heavy breed there are several to choose

from: Cochins, Croad Langshans, Dorkings, Jersey Giants, Orpingtons, Plymouth Rocks and Wyandottes, and of all these, the Wyan-dottes are likely to win most major awards at British shows, and perhaps White Plymouth Rocks at American shows.

There are several breeds that were once both very popular and regarded as living works of art because of the expertise necessary to produce good specimens; Minorcas and Rosecombs are two such examples, and judges will usually reward their exhibitors for their particular skill.

White and black breeds win more major awards than patterned varieties, mainly because the patterns are more difficult to breed to perfection. However, they are in less demand than the patterns by buyers of pure breeds who just want them as garden pets, and will not be showing. There are several more breeds, as detailed below, that regularly win major awards, but are not very popular with pet buy-ers. So although very successful exhibitors are able to charge high prices for their champion strain, this is a limited market; furthermore, a new breeder of (say) Black Old English Game bantams will take several years (at least) before they become anywhere near as well known as the existing experts.

Localiz ed Popularity of Breeds

After visiting a few shows in your area, it will be clear which varieties are bred locally, and whether this is by just one or two breeders, or by a larger group of enthusiasts. Assuming the new fancier already has a mental shortlist of likely breeds, it has to be decided if it is a good idea to join the local enthusiasts, or whether it might be better to try something different. Thus from the white breeds example above, if there are already White Wyandotte exhibitors in your area, by joining them you will have the advantage of more interesting and competitive shows, but possibly limited sales opportunities – and if there are several experienced breeders already, of course it will be more difficult to win. More major awards might be won if another breed is chosen, a variety that regularly wins top prizes, but is not currently bred in your area. This policy has the additional advantage of two or three years' good stock sales with no local competition.

UTILITY PURE BREEDS

Most pure breeds, even the really fancy types, will lay enough eggs for most families, but only a few are popularly thought of, and sold as, utility breeds. In the UK the main three are British-type (tailed and crested) Araucanas, Cuckoo Marans and Welsummers; in the USA, Ameraucanas replace British Araucanas. To this list might be added the Autosexing Breeds, Barnevelders, Croad Langshans, French Marans and New Hampshires. The very dark brown egg-laying Penedesencas from the Catalan region of Spain seem likely to become another popular breed in this group.

All of these can be successfully exhibited (in the UK) except for French Marans and Penedesencas, which are (in 2004) not standardized. However, these utility breeds have not won as many 'Best in Show' as other exhibition breeds. Nevertheless, surplus stock is very easy to sell, and trios of Welsummers have reached very high prices at pure breed poultry auctions.

There are distinct utility strains of Light Sussex, Rhode Island Reds and White Wyandottes, all different from the exhibition types of their respective breeds. These strains are a significant part of our poultry heritage, but were, by definition, never intended to be shown. Some shows have utility classes, which might help sales if a lot of public were attending, but this is not quite 'exhibition poultry keeping'. The breeders of these strains may wish to form their own clubs and events to promote this sector of poultry breed conservation.

Breed Clubs and the Other Fanciers

Most, hopefully all, of the breed clubs have efficient secretaries who promptly answer correspondence, publish interesting newsletters and fulfil all the other necessary duties. They should be fully national bodies, although the smaller clubs are bound to have large areas with no members at all. This is bad enough in the UK, but imagine some of the distances between members of a breed club with a hundred members in Australia or the USA; even if it is intended to buy in a breed not currently in your area, how far will you wish to travel?

Personalities are also a factor. We are all aware of stereotypes about types of people and breeds of dogs and the same is true, to a lesser extent, of people and breeds of poultry. It has never been recorded, but it is certain that many fanciers have started to keep, or stopped keeping a breed because they liked, or didn't like, the other people with that breed.

Many fanciers choose to keep several colour varieties of one breed, or the large fowl and bantam versions of one breed. Some can be tempted to go for every standardized variety, which in many breeds would be too many to manage. This approach is very good for the conservation and improvement of rare colour varieties, as the breeder will have suitable stock for crossing varieties within the breed where necessary.

As regards poultry shows, these are divided into sections, each with several breeds. The groupings are different in the UK from those in the USA, but fanciers in both countries might decide to maximize their chances of winning major awards by having one breed in each of several sections. Thus a British bantam fancier could choose one hard feather breed, one heavy breed, one light breed, one true bantam breed and one rare breed.

A BRIEF SUMMARY OF THE BREEDS

This book is not intended to be an encyclopedia of poultry breeds, or a replacement for a book of standards, so the following paragraphs comprise just a few comments that might help novice fanciers decide which breed(s) to keep. They are grouped according to the sections in British poultry shows.

Large Hard Feather

Despite having been illegal since 1849, cockfighting continues in secret. Some of those involved breed their own birds for this purpose,

17

Grey Carlisle-type Old English Game, large male.

Grey Oxford-type Old English Game, large male.

but others steal them from exhibition breeders. These exhibitors are naturally reluctant to enter shows that have catalogues, because they do not wish to have their address published. Many have given up after repeated thefts, however enough remain for this to be a very competitive section.

Carlisle-type Old English Game: This broad-bodied type regularly wins major awards, especially in the north of England, as the name suggests. It is (the author believes) unknown in the USA, Australia or New Zealand. There have been many decades of argument as to the extent to which Carlisles have changed from the historic type, and the ideal bodyweight for showing. Many adult cocks are rather inactive, and are therefore becoming less attractive to potential thieves.

Oxford-type Old English Game: A smaller-bodied, more active type, similar to the OEG as bred in the USA, Australia and New Zealand. Some UK shows cater for both types, other areas seem to have either all Carlisle or all Oxford breeders. In these areas it is best to keep the same as everyone else.

Indian Game (Cornish in the USA): These are not as popular as the OEG, except perhaps in their traditional stronghold, Cornwall. A good Indian is just as likely to win 'Best Large Hard Feather' at a (UK) show as a Carlisle, and has a rather better chance than any other breed in this group. They are difficult to breed, and because they have rather brittle plumage, they are equally difficult to keep in show condition. Nevertheless, quality Indians change hands for high prices, which is another reason to consider them.

Modern Game: Large Moderns are very rare, maintained by a few historically minded enthusiasts. They are hardly ever as large as they should be, so are not the prominent show breed they once were. Experienced breeders might be able to make further progress in their conservation.

Jubilee Indian Game, large male. This is the same pattern as Dark Indian Game, but with white replacing the black. The nearest American equivalent, the White-Laced Red Cornish, is a different pattern.

Asian Hard Feather Breeds (Asil, Malay, Shamo, etc.): These have been enjoying a rapid rise in popularity in the UK since 1990. Some of our more conservative judges positively dislike them, but Asians were beginning to win more top awards at the time of writing; for example, a Shamo pullet become Supreme Champion at the 2002 National Championship Show for Peter Campbell.

Hard Feather Bantams

Old English Game Bantams: These are by far the most popular breed at British and American shows, although the breed is very different in the two countries. In the UK, OEG bantams have broad bodies and rather small, narrow tails, and judging is almost entirely on body shape, with details of plumage colour being ignored. American OEG bantams have large fanned tails, and plumage colour and pattern is closely examined. The author has not had direct contact with American OEGs, but doubts if they feel as muscular as British judges would expect. Their popularity is not uniform, with rather fewer in the south-east of England, for example; at the other extreme, at some northern shows there can be as many OEG bantams as all other breeds combined. There are also miniature Oxford-type OEG, but these are very rare.

Indian Game Bantams: As with the large version, perfect exhibition Indian Game bantams are difficult to produce, but will win prizes for those who like them and have the skills. Note that American shows are divided into different sections, and 'Cornish' are in the 'All Other Combs, Clean-Legged Class' (UK 'Section' = US 'Class'), which they also appear to win regularly. Breeders should be aware that a British Jubilee Indian Game is not the same pattern as an American White-Laced Red Cornish.

Modern Game Bantams: These tiny, elegant birds have been a favourite with exhibitors and judges in most poultry-showing countries for over a century. It is to be hoped that this continues to be the case, although the author doubts it because not many new breeders are taking them up.

Asian Hard Feather Breeds: Ko-Shamo are rising in popularity so fast in Britain, Germany and elsewhere in Europe, that they seem set to become one of the most numerous breeds of all. They are small, easy to keep, and the cockerels do not crow very much. Related breeds include Chibi-Shamo, Tosa-Chibi (both regarded by some authorities as sub-varieties of Ko-Shamo), Nankin-Shamo, Tuzo and Yamato Gunkei. These are much rarer, but are likely to attract more interest in the wake of the rise of Ko-Shamo. Malay bantams were once a popular

Shamo, large male. This Japanese breed was virtually unknown in Europe before 1975, but has attracted many European breeders in the thirty years since then.

Jubilee Indian Game, bantam female. Those who are familiar with the White-Laced Red Cornish will notice that female Jubilees look similar at a distance.

exhibition breed in Britain, but that was a long time ago, 1880 to 1914; since then they have been much rarer, and of very variable quality, in Britain at least. Malay bantams have enjoyed more support, and are more consistent, in Germany, but it will need several years of effort by more serious breeders, certainly more than there is at present (2005), if Malay bantams are ever again to win major awards regularly.

True Bantams
This is the section, in British shows, for those breeds that are not miniatures of large breeds. Pekin bantams are included in this section because our standard type, with its very low carriage, is considered different enough for them not to be regarded as miniature Cochins, which is the case in most other countries.

Barbu d'Anvers (= US Antwerp Belgians), Barbu d'Uccles, Dutch, Japanese (Chabo), Pekins, Rosecombs and Sebrights are all very popular, and all regularly win 'Best in Show'.

Rosecombs and Sebrights are, however, too difficult for most novice fanciers, unless an experienced breeder is ready and willing to advise.

Barbu d'Everberg, Barbu du Grubbe and Barbu de Watermael are all much rarer, but are not classed as 'rare breeds' here because they are covered by the Belgian Bantam Club, and not the Rare Poultry Society. The RPS does cover Booteds, Deutsche Zwerghühner, Nankins and Ohikis, so these are in the 'rare breeds' section. There are some additional true bantam breeds that are not bred in Britain at all, French Pictaves for example; if imported, they would become 'rare breeds' here until (if ever) they became sufficiently popular for a separate breed club to be formed.

Large Light Breeds
The large soft feather breeds are only formally divided into 'light' and 'heavy' at the National Show, but some local shows have cups for 'Best Light Breed' or 'Best Mediterranean',

White Pekin bantam female. This was show champion for Anne Peutherer at Reading Bantam Club's Show, the UK's largest all-bantam show. It is a clear demonstration of the exhibitor's skill to produce such profuse and pure white plumage. Note that the British Standard Pekin bantams should have lower carriage than Cochin bantams, as shown elsewhere; there is not always much difference.

Black Mottle Barbu d'Uccles bantam male. The foot feathering is more brittle on this breed than it is on Pekins, making Barbu d'Uccles of all colours more difficult to keep in showable condition. It is (sadly) all too obvious which birds have been entered by experts, and which by novices.

usually donated by a keen breeder some time ago. Leghorns, Minorcas, Polands and Silkies are the light breeds most likely to win major awards, and of these the Polands and Silkies are the most saleable to non-showing poultry buyers. Araucanas and Welsummers, as mentioned in 'utility breeds' above, are easy to sell but less likely to win 'Best in Show' or other major awards.

The remaining breeds in this (UK shows) group are Anconas, Derbyshire Redcaps, Hamburghs, Scots Dumpies and Scots Greys. In many areas they will have to compete in mixed 'AOV Light Breed' classes, and the result may depend on whether the judge likes the breed or not. Excellent condition and presentation is essential for success in such mixed classes, and self-coloured, or those with a fairly simple pattern, have an advantage over 'difficult' patterns. On this basis, Black Hamburghs and Scots Greys are likely to do well, and despite the difficulty of obtaining perfect spotting, a lot of older judges like Anconas because they fondly remember them from when they were a popular commercial breed.

Light Breed Bantams

Much the same considerations apply as for large light breeds, except that there are many more of them shown, especially Ancona and Hamburgh bantams. In the case of Leghorn bantams, only three of the colour varieties (blacks, browns and whites) are regular prize winners. Minorca bantams are more viable as an exhibition breed than large Minorcas because of the costs involved in this short show-career variety.

Large Heavy Breeds

Large Orpingtons, Sussex and Wyandottes attract a lot of breeders, both novices and old hands. Australorps, Croad Langshans and New Hampshires are probably easier for new fanciers to breed and prepare for shows.

Ancona bantam male. The neat white spots, straight comb and good white ear lobes only last for one season on most Anconas, so a fresh show team must be bred annually. Most Ancona bantams in the UK have rather small tails in proportion to the tails of large Anconas.

Light Sussex, large male. These have won 'Best Large Soft Feather' many times at British poultry shows. Many were once reared in orchards, the trees providing the shade needed to prevent the white plumage going 'brassy'.

Marans, as said in the utility breeds section above, are easily sold, but do not win much. Brahmas, Cochins, Faverolles and Frizzles all require expert management. Barnevelders are easy to keep, but only a limited proportion have the correct colour and markings. Dorkings seldom look as good in reality as they were depicted by the famous specialist poultry artists, possibly a result of inbreeding. Large Plymouth Rocks are very rare in Britain, and would be a good choice for someone prepared to make a serious effort. Rhode Island Reds are a favourite among the generation of judges who are old enough to remember them when they were kept on almost every farm; younger judges will only see them as they are now, which is often with very poor feather quality.

Heavy Breed Bantams

The same remarks apply as to the large heavy breeds. Australorp bantams are shown in much greater numbers, and win top awards much more than Large Australorps. Conversely, more excellent Large Brahmas and Croad Langshans are seen than bantams. Dorking bantams are very rare indeed. Sussex and Wyandotte bantams are two of the most popular showing breeds of all, and the less-than-perfect patterned varieties are easily sold as pets.

Rare Breeds

Many rare breeds are, by definition, very seldom seen, and because of the small numbers bred are not very good. Even when good specimens are exhibited, some judges are not familiar with all the breed standards. The rare breeds most likely to win major awards are large and bantam Spanish, Sultans, Sumatras, Yokohamas and bantam German Langshans and Rumpless Game. These are shown regularly and are easy to judge by non-specialists.

Wheaten Rumpless Game bantam female. This breed is associated with the Isle of Man, also home of the tailless Manx cat. Because 'Rumpies' are judged on a very similar basis to the popular Old English Game bantams, non-rare-breed specialist judges find them easier to assess than some other rare breeds.

Fawn and white Indian Runner drake. Runners of all colours are comparatively easy to assess by non-waterfowl specialist judges, as often happens at smaller events where two or three judges have to judge the whole show. Call ducks are also fairly easy to judge by non-specialists, hence these are the two most popular breeds of duck for showing.

Ducks, Geese and Turkeys

This book is primarily concerned with breeding and showing chickens and bantams, but many exhibitors have ducks, geese and turkeys as well, so they can be considered here. All three species require a lot of space, which suburban poultry keepers seldom have; even a small flock of Call ducks will transform a tidy garden into a smelly, muddy disaster area.

For those with the space, Call ducks and Indian Runners are the most competitively shown. Muscovy ducks are often thought of more as a utility table breed, but a big Muscovy drake will impress most judges. Aylesburies are rather rare, being inbred and difficult to breed, but this historic breed urgently needs the support of anyone with the enthusiasm to overcome the problems. Read Chris and Mike Ashton's excellent book *The Domestic Duck* (also published by Crowood) for full details of all breeds.

The most striking-looking breeds of geese are Africans, Chinese, Sebastopols, Steinbachers and Toulouse. At many small- to medium-sized shows there will not be a specialist goose judge, and there are often only mixed classes ('AV Gander' and 'AV Goose'), so these breeds are likely to win over conventional-looking geese, the finer points of which will not be appreciated by a general poultry judge whose waterfowl knowledge is limited to the popular duck breeds.

The Turkey Club UK has started to increase interest in these magnificent birds, but at the time of writing (2004) it was too early to say how far this could go. Turkeys also need quite a lot of space, but they do not make a mess like waterfowl, and are less noisy than cockerels or Call ducks. The space they need when transported to shows is obviously much greater than that for bantams, but suitable trailers can be bought at reasonable prices, and many country people have vans, pick-up trucks or horse trailers. The problem of non-specialist judges at most shows is even more acute with turkeys than geese. As with mixed classes of geese and AOV chicken breed classes, good presentation is essential for success, and in most cases the biggest turkey with the cleanest feet, straightest toes, best plumage condition and without breast blisters will win.

ABOVE: *Sebastopol gander. This breed needs considerable skill to present with clean plumage in good condition. However, it can be worth the effort, as a good one will really stand out when the judge is deciding which 'Best of Breed' will go on to win the 'Best Waterfowl' award.*

Cröllwitzer turkey stag. Since first being imported to the UK during the 1970s, these have become a favourite among those with enough space to keep some fancy turkeys.

CHAPTER THREE

The Organization of Poultry Showing

This chapter is in two main sections: the first describes the general structure of the poultry-showing world; and the second, how a new local poultry club would be formed in Britain and go on to hold its first show. A similar process, adapted to suit relevant national rules, could be applied in other countries. There are many breeders interested in old poultry breeds around the world who have no clubs to join, and the author sincerely hopes this chapter will encourage the formation of some new clubs somewhere.

THE POULTRY-SHOWING WORLD

(*See* also the appendix at the back of this book for an extensive, but probably not complete, list of poultry-showing organizations; this is intended as a guide to the general strength of our hobby worldwide.)

National Poultry Clubs of Great Britain

In Britain, the fancy is governed by the Poultry Club of Great Britain (PCGB), founded on 14 November 1877 at the Crystal Palace Show. There had been a previous attempt at a national governing body in 1863, but it fell apart in acrimony between the leading figures of the day after only three years. During these three years the first Poultry Club published the first-ever book of breed standards, *The Standard of Excellence In Exhibition Poultry*, a small, fifty-six-

page volume that appeared in 1865. It was mainly the work of Messrs Dixon, Teebay and Tegetmeier, all well known names to those interested in the history of poultry breeds. During the 150 or so years since then, there have been numerous books of poultry breed standards published in several countries and languages, each edition larger than its predecessor.

Other functions of the PCGB and its equivalents elsewhere are administering the judges' exams, issuing closed leg rings, and representing the hobby to government departments and the public. Also, the PCGB publishes the *British Poultry Standards*, and because those standards have originated from the breed clubs, there are sometimes disagreements to be reconciled. Although it might be reasonable to assume that the breed club membership will be the experts, sometimes a broader view has advantages. The PCGB organizes the National Show, which normally has 5,500 entries.

Affiliated to the PCGB are over 130 local poultry clubs (including the poultry sections of the agricultural shows), over fifty breed clubs, and a number of 'associate' clubs. The associate clubs include some that are similar in function to the breed clubs, but are larger and cover several breeds, for example the British Waterfowl Association, the Domestic Waterfowl Club, the Goose Club, the Rare Poultry Society and the Turkey Club UK. There is also the Utility Poultry Breeders Association, which encourages those who wish to record the egg production and growth-rate data of any pure breed. Since the commercial poultry industry switched to using hybrids there have always been some breeders more interested in maintaining the original productivity of pure breeds than in showing them. It is to be hoped that they help to make the UPBA a successful organization for utility breeders.

Clubs in the USA

In the USA there are two governing bodies, the American Poultry Association (APA) and the American Bantam Association (ABA). A lot of Canadian fanciers join one or both of these. The UK once had a similar arrangement, with a separate British Bantam Association existing from 1933 until 1952, when it amalgamated with the PCGB. Poultry shows were once an advertising medium for commercial poultry breeders, and as a result the bantam enthusiasts felt they did not get fair treatment in terms of judges, classes and prize monies from show organizers. Thus they formed separate bantam clubs because they felt aggrieved. Since the 1960s, when the shows have been purely a hobby, and bantams have usually outnumbered the combined total entry of large fowl, ducks, geese and turkeys at most shows, there has been less need for separate bantam clubs.

The APA was formed at Buffalo, New York, in 1873, making it the oldest continually running national poultry club. The ABA followed in 1914, founded by a group of breeders in the Chicago area. It has a much more liberal attitude to new colour varieties and imported breeds than the APA. The separate ABA *Bantam Standard* books – there have been several editions since 1965 – have many more colour varieties for some breeds than appear in the APA standards for the same breed. The author does not know how much, if any, argument there is between the two bodies over these differences. Publicly, '... continued co-operation and friendship between the APA and ABA ...' is usually emphasized.

Affiliated to the APA and ABA are many breed clubs and local poultry clubs, as in the UK. Because of the sheer size of the USA and Canada, state-level organizations are very important for day-to-day running of the fancy.

Clubs in Australia and New Zealand

On the other side of the world, fanciers in Australia and New Zealand also have to cope with huge distances, or in the case of New Zealand, a North Island, South Island split. The Australian poultry show scene is also largely run at state level. The New South Wales Exhibition Poultry Association has about fifty local poultry clubs affiliated to it, which between them can enter over 3,500 birds at the main NSW state show. There are also large state shows at Melbourne (Victoria), Adelaide (South Australia) and Brisbane (Queensland).

Clubs in Europe

Distance is not a problem in the countries with the greatest concentration of show poultry fanciers: Belgium, Denmark, France, Germany, Luxembourg and The Netherlands. In these countries most of the shows are very large events and include fancy pigeons, rabbits and other small furry animals, sometimes cage-birds as well. Some of the national governing bodies and local show-organizing clubs cover two or more types of small livestock. This was once commonly the case in the UK as well, where they were called 'Fur & Feather Clubs'.

In Germany and The Netherlands the emphasis is very much on large shows, usually held over at least three days. The 78th German National Show in 1997 had a total of 28,361 exhibits from about 3,500 exhibitors; these included 135 turkeys, 37 guinea fowl, 484 geese, 1,106 ducks, 4,676 large fowl, 8,118 bantams, 11,250 fancy pigeons and a further 2,617 birds of all types in the junior exhibitors' section. In addition to trio (one male, two female) classes, German shows also have 'Volieren' (one male, six female) classes. Only the most expert and specialized breeders are likely to have six matching females, especially of patterned varieties.

Denmark has a relatively small human population, about 5.4 million, so a membership of approximately 4,300 in the Danish National Poultry Club (DFfR) and an entry of 4,318 poultry in their 2001 National Show indicates that our hobby is comparatively strong there.

BREED CLUBS

These are for all the breeders in a country of any one specific breed. Some of these clubs are for a group of related breeds: for example, the British Belgian Bantam Club covers five breeds from Belgium. In Britain there is the Rare Poultry Society to cater for breeds without a club of their own, with the similar Society for Protection of Poultry Antiquities (SPPA) in the USA. Some breeds have more than one club in a country, either for specific colour varieties or for a region of the country. The separate colour clubs are often a legacy of an ancient argument over the desirability of a new colour variety being standardized, or the dominance of one colour in winning major awards, especially self-white birds versus patterned varieties.

The main activity of a breed club is to publish newsletters, year books and breeders' lists. Most of their annual shows and meetings are at national poultry shows, although a few make alternative arrangements – not always popular. For instance, the (British) Indian Game Club Show has been held in Cornwall for over a century, which may appeal to traditionalists, but is not very inviting for potential new members in the north of England.

Breed club members also have regional shows in which to compete, and local shows act as hosts for these. They do so partly for prestige, partly because local club committee members are also likely to be breed club committee members, and partly because the local clubs always appreciate the extra entries. Exhibitors will travel outside their normal area for a breed club regional, and may also keep other breeds that will add entries to the other classes.

International Organizations

The most important by far of the international organizations is the European Poultry and Rabbit Breeding Entente, formed in 1938, although World War II prevented any activities until 1948; it is also for fancy pigeon breeders. Fantastic shows are held, attracting exhibitors from all over Europe, and staged in different countries each time (subject, of course, to there being no disease control regulations preventing this, which has always been the case in Britain). The PCGB was beginning to get involved at the time of writing, but this seemed likely to be very controversial with some breed clubs, namely those breeds where the birds exhibited in Britain are noticeably different from the birds in existing Entente countries.

Nevertheless, most breed clubs have at least some contacts with equivalent clubs in other countries, which in a few cases has progressed to a more formal international breed club. Chabo, Poland, Brahma and Araucana breeders seem to be leading the way. In the 1970s and 1980s an American breeder, Gerald Franklin Wright, wrote to fanciers all over the world as the founder of a whole range of 'international' breed clubs, though none of them ever became a reality. But he was not the only one with this appealing idea. Potential members should

A successful show held by the Midsomer Norton Club in the Showering's building on the Bath and West Agricultural Showground, Somerset. The permanent buildings on agricultural showgrounds, and indoor riding school buildings, may be easier to hire for poultry shows than village halls and other venues where the potential mess may put off hall owners.

check with other breeders they know personally before they get too involved with 'international' clubs, to check if they are real. Remembering that fanciers in Belgium, Denmark, Germany and The Netherlands are often in regular contact anyway, any 'international' clubs based in these countries should be genuine groups of like-minded breeders.

Even technophobic chicken keepers have computers now, so international contacts are sure to become much more commonplace. There are thousands of poultry-related web sites, which should be particularly helpful for geographically isolated breeders.

LOCAL POULTRY CLUBS

Active local clubs are vital for the future of our hobby; the best of them have regular monthly meetings as well as shows. Not all do so; they may start with meetings that are abandoned as interest fades. It is difficult to arrange guest speakers or other activities to appeal both to beginners and to long-time poultry keepers. Also, some new members do not appreciate that these clubs are very show-based and local smallholders' societies may be more suitable for utility-minded domestic poultry keepers.

This section describes the procedure for forming a club and running shows in Britain and Ireland, and is mainly intended for new fanciers who intend to take an active part in an existing club. However, the author sincerely hopes that it may also help the founding members of new poultry clubs in areas without a club at present. For instance, at the time of writing, there were some shows in the Republic of Ireland, but there could have been a lot more.

Apart from meetings, local clubs typically organize at least two shows a year. One will be the main winter show, held in a village hall or some other suitable venue – though it is becoming more difficult to find halls that will take poultry shows, or any other livestock-related activity. Some shows now use an indoor riding school as a venue, because any remaining wood shavings or other mess is not a problem in such buildings. In the summer, local clubs usually organize the poultry section of agricultural shows, 'country shows' and similar events. Chickens do not look at their best in the summer because they are in the breeding pens and plumage is becoming tatty. A local club committee – effectively all the leading fanciers in the area – may realize that they do not have enough presentable birds between them to stage a proper competitive show; however, they appreciate the importance of publicity, and so put on displays wherever they are invited to do so. Fêtes, and open days at 'rural interest' establishments, have been the venues for successful local poultry club displays. They always attract a lot of public interest, especially in towns and cities where many children have never seen a live chicken or turkey close up before. And if the public are happy, the organizers will be happy, and out of a large crowd there will usually be a few interested enough to buy some stock or even join the club.

A new club will probably be the brainchild of a group of existing exhibitors who are inspired because a few more keen new fanciers have started up. Furthermore, there are bound to be some pure breed poultry keepers in the area who would join if approached, and local animal feed suppliers may know of even more potential members.

The next step is an inaugural meeting, probably at someone's house; at this stage, there will probably be no more than ten people. The election of officers is seldom needed in reality, as the new club will need all the volunteers it can get; it is simply a matter of agreeing who does what, preferably with at least two main organizers. Many clubs have separate 'general secretary' and 'show secretary' posts to spread the workload, and there will also be a president and a chairman. In the UK, the president is an honorary post for respected fanciers or financial donors, and it is the chairman who should be the real group leader. It is to be hoped that all the remaining committee members will bring their own special contribution to the success of the club: the use of a barn to store the cages, a truck to transport them, office/printing services or show catering.

RUNNING A SHOW
Allocation of Funds
Money will be needed for the following:

- Printing show schedules and entry forms.
- Postage, to send out the above.
- The hire of the show venue (village hall, riding school). Many new clubs have tried holding shows in farm buildings, if these are available for free, but few are really suitable as they seldom have adequate toilets or catering facilities. There might also be problems with event insurance on farms. Remember to book your hall/school for the day/evening before the show, to allow enough time for putting up cages. This takes several hours, so cannot be done early in the morning of the show.
- Affiliation to PCGB, and public liability insurance for poultry shows available from PCGB. The PCGB will advise on current regulations concerning poultry health, vaccinations and so on.
- Show cages, trestles and staging (plywood sheets). Eventually the club will need to buy a complete stock of cages, but it may be possible to hire them from another club, for the first few shows at least. The trestles will be made by club members, and standard sheets of plywood can be machine-sawn to make two staging sections, one for large fowl cages, one for bantam cages. Very large pens for geese and turkeys are usually homemade. A new club should certainly aim to buy enough show cages, staging and trestles to put on modest public displays as soon as possible. Some existing clubs may be willing to sell some of their older cages.
- The purchase of rosettes, prize cards and perforated judging forms.
- Cups, cash prizes or other prizes for Show Champion, Reserve Champion and section winners. These are often donated.
- Refreshments for judges (who will expect to be fed for free) and exhibitors (who can be charged, to recover expenses).
- The travelling expenses of judges.
- Plates for egg classes, and bales of wood shavings for show cages.

There will be income from show entry fees, and most clubs hold a raffle to raise a little more cash. The show budget should cover immediate expenses, with a profit for the show cage fund. A new club will obviously need to raise a substantial amount before its first show.

Pre-Show Meeting
The first pre-show meeting should be held at least three months before the intended show date. The main matters to be arranged are the date of the show, the judges, the venue, the purchase and transport of cages and equipment, and catering.

Date of show: Subject to venue availability, it should not clash with any other local show within at least 100 miles, or any of the major shows.

The judges: A new club should hope for an entry of about 250 birds, which will need two judges. Ideally choose two who live near to each other and who could travel together, and not so far away that they will need huge travelling expenses. Any qualified judges who live very close will hopefully be on the club committee, and will be showing.

The venue: Confirm all bookings, and make sure the details are settled. Summer shows are usually in tents, and are often part of a larger event (an agricultural show or similar). One of the poultry club committee will probably attend meetings of the larger event to arrange all the details.

Purchase and transport of cages and other equipment: It is to be hoped that the poultry club will have several members with trucks, horseboxes or large vans who are willing to transport everything from wherever it is all stored to and from the show. Some bales of wood shavings (US = woodchips) will be needed for the cages and egg show plates and paper plates for the egg section exhibits. Water pots will be needed for the birds, although some exhibitors will bring their own. For the rest, buy some packs of cheap clothes-pegs and some vending-machine plastic cups; empty (and washed) cream/yoghurt pots can also be collected by members for this purpose, and these can be pegged to the cage wire. Cage ends and tops will need to be secured, and freezer-bag or garden 'ties' are best for this.

Catering: Not much is needed at summer agricultural shows because there will be a lot of catering outlets at the show; therefore only a selection of hot and cold drinks need be provided. At winter shows in a hired hall there is much more to be done, and several people will be needed to buy, cook and serve the food that cold and hungry exhibitors expect. The committee will probably have attended enough shows to know the level of catering generally provided.

Most clubs will hold a second meeting to finalize arrangements.

The Secretary's Duties
The show secretary has various duties that include applying for breed club 'specials', booking the judges, printing and sending out entry forms, and processing entries. He/she is also responsible for setting up the show, its smooth running during the day, and organizing the clearing up afterwards.

Breed club 'specials' (prize cards or rosettes): Some of the breed clubs require a minimum number of classes for their breed, so it is best to check on their requirements before the schedule is printed. New clubs should start with just the breed clubs that they are confident will have a reasonable entry. Other committee members can help by getting specials from the clubs they are members of.

Booking the judges: The secretary should phone prospective judges initially, and if they agree, then write to confirm.

Printing the schedule: All relevant details must be included: the judges, the venue, start and finish times, the closing date for entries, and other details. The classes provided – at British shows, other countries do not have a pre-set schedule of classes – will be different for each show, reflecting the breeds kept locally. All schedules should follow the normal show sections, and there should be at least one AOV class for each section. Some exhibitors could be permanently put off ever entering your club shows if there is not even a suitable AOV class for their breed.

Two sections that are regularly poorly catered for in the UK are large hard-feather and rare breeds. In the first case, many shows only provide classes for Carlisle-type OEG and Indian Game, forgetting (?) about Oxford OEG, large Modern Game and the Asian hard-feather breeds. In the case of rare breeds, some clubs seem to forget that this is now a section in its own right, and that Sumatra and Yokohama males need extra-large cages for their tails. Show organizers should also remember that if one or more classes are to be provided for non-standardized colours of non-rare breeds (for example, 'Coronation Sussex'), these should not be in the Rare Breeds section. Some clubs simply do not have space for turkeys or geese.

British fanciers prefer to remain with their traditional show class system, and the inevitable mixed ('AOV') classes because they are competitive, and so if they are the only exhibitor of their breed, they would rather take their chances with other, similar breeds than win by default.

Show secretaries should keep up to date with changes in breeds kept in the area, and should be willing to provide additional classes when needed. Breeds can also decline in popularity. Thirty years ago the large fowl classification given in the example opposite would have included several for specific colours of Old English Game (Carlisles in some areas, Oxfords in others) virtually everywhere in the UK; but while this is still the case in some shows, it no longer applies everywhere. Some show officials have resisted changing classifications in breed popularity because their own favourites are

---

Example of Large Fowl Classes for a Small British Show

01) Large Trio	09) AOV Rare Breed M	17) Marans M/F
02) AC Carlisle OEG M	10) AOV Rare Breed F	18) Buff Orpington M/F
03) AC Carlisle OEG F	11) AC Araucana M/F	19) AOC Orpington M/F
04) AC Oxford OEG M	12) White Silkie M/F	20) Rhode Island Red M/F
05) AC Oxford OEG F	13) AOC Silkie M/F	21) Light Sussex M/F
06) AOV Large Hard Feather M	14) Welsummer M/F	22) AOC Sussex M/F
07) AOV Large Hard Feather F	15) AOV Light Breed M	23) AOV Heavy Breed M
08) Sumatra or Yokohama M/F	16) AOV Light Breed F	24) AOV Heavy Breed F

In this example Sumatras and Yokohamas are separate from the other rare breeds so they can be given double-sized cages, and the Orpingtons are divided into 'Buffs' and 'AOC' because of the separate breed clubs. There will probably also be some classes for generally rare or minority breeds, Derbyshire Redcaps perhaps, if there are several breeders locally.

Example of Form for Collating Entries
Class 02, AC Carlisle OEGM

Cage No.	Exhibitor's name	Details of exhibit	
	A Adams	Black-red	It is hoped that all exhibitors
	B Brown	Brown-red	do actually state the colour of
	C Carter	Duckwing	each bird.
	D Davis	Pile	

The 'cage number' column cannot be filled in until all entries have been listed. This is because the numbering follows through all the classes, as Class 01) 1–6, 02) 7–11, 03) 12–19 and so on. When all entries are in, cage numbers are added.

---

suffering a decline. However, it will not help revive one breed's fortunes by trying to hold back another.

Preparing the mailing list and sending schedules with entry forms: Other clubs in the region might provide copies of their members' lists to help build a mailing list of potential exhibitors. Keep show catalogues with addresses for further names. Additional entries can be attracted by placing schedules with entry forms at animal feed retail outlets in your area.

Processing entries: Another of the secretary's duties is to prepare forms, by hand or on a computer, to process entries. The entries have to be arranged into the breed classes. There may be an unexpectedly large number of entries in some classes, which the secretary might wish to divide into separate classes for males and females, or for specific colour varieties. For example, the AOC Silkie M/F class in the example above might be divided into three: AOCM, Black F, AOCF; and the secretary must decide if, or how much, to revise future schedules.

As far as possible the order of classes in the schedule should require the minimum number of changes in cage size; however, some changes are inevitable. The waterfowl section will probably include extra-large goose pens, large fowl cages (for most duck breeds) and bantam cages (for Calls and other small breeds).

Once this stage is completed, the secretary will need to provide a 'penning slip' for each exhibitor and a cage plan for when the show is set up. A lot of show secretaries simply fill the cage numbers in the 'sec's use' column on the entry form, and then give these forms back to each exhibitor as they arrive on show day. Those show secretaries who use their computer to do the administration print out separate penning forms, for which suitable programs are available.

Show secretaries are required to keep a permanent record of the names and addresses of all exhibitors in case there in a poultry disease outbreak in the area.

A set of numbered labels is needed to fix on the show cages. Regular exhibitors will have seen several types around the shows. The easiest way for a new club to produce these labels is to buy a book of cloakroom/raffle tickets for the numbers and some packets of parcel labels to stick them on: the type of labels with small clips are ideal, to hang on the cages. Paper plates are needed for egg classes, and a bale of shavings for the cage floors.

Setting up the show: There will hopefully be enough helpers to set up the show cages to leave the secretary as 'The Boss' on the day/evening before the show. It is essential that the secretary has a clear idea of where all the blocks of goose/turkey, trio/long-tail, large fowl and bantam cages are to be erected. Try to avoid any classes being split in distant parts of the hall. If space and the number of cages available permits, erect a separate 'Champions Row' in a prominent place and a few spare pens in case of mistakes, or difficulties in making the classes fit the banks of cages. Other tasks include catering arrangements and egg section plates.

On the morning of the show: The penning slips (cage numbers for each exhibitor) are laid out on a table and the catering should be in full swing as early as possible. The show secretary and at least two helpers should be free of other duties so that they are available to cope with any problems that arise. It would probably be a good idea if the secretary's mobile phone number is printed on the schedule, or at least known to the judges, so that the show organizers can

be informed of any car breakdowns or other problems occurring to people on their way.

The event will run much more smoothly if the secretary has some helpers to write out the prize cards so that he/she is free to resolve any problems. Judges are provided with a free lunch, which is either prepared by the show caterers, or they are taken for a pub lunch. Organizers and judges aim to complete the class judging before lunch, going back to choose section winners and the Show Champion afterwards. As British show prize cards do not include 'judges' comments', it is expected that judges will stay at the show for an hour or so after all judging is finished to discuss the results with any unhappy exhibitors. At the end of the show, someone has to draw the raffle prizes, present the trophies and then formally announce that it is time to go home.

After the show: Those exhibitors who live fairly close to the show should resist the temptation to rush home, as all the cages have to be taken down, loaded and returned to wherever they are stored. Then the hall has to be swept clean. It is seldom quite clean enough, so it is a good idea to give the caretaker a generous tip in advance.

Details of the main winners should be sent to all the relevant magazines: *Feathered World*, *Fancy Fowl* and *Practical Poultry*. Both the name of each exhibitor and the full variety description of each winning bird should be sent as well as names of the judges. If a few lines of additional news can be added to make the results page more interesting, the editor will be more than pleased. It is to be hoped that this will be good news, perhaps the Show Champion going to a junior exhibitor. It might be more dramatic news, such as severe weather conditions – freak summer storms have brought the tent down at a few agricultural shows. Good quality photos of the winning birds, with or without their owners (or the wrecked tent!) will be welcomed.

Finally, there should be a committee meeting to review the show's successes and failures in the hope that the next show will be even better.

CHAPTER FOUR

Housing and Equipment

Compared with small-scale poultry keepers who primarily concentrate on egg production, exhibition breeders have much more housing for any given number of birds. Growing stock will be housed in larger-sized groups, the number in each batch being determined by how many were hatched, but adult exhibition stock will normally by kept in small groups. Fifty layers can easily be kept in one group. A 10 × 20ft (3 × 6m) shed would be spacious enough, even if they were not allowed access to the great outdoors. Fifty exhibition bantams could be kept in a shed the same size, but the shed would be divided into several pens, some of which would typically be raised in two or three tiers to get greater capacity in the building. Anyone breeding the large exhibition varieties, Brahmas or Orpingtons perhaps, will need two, possibly three, sheds this size for fifty birds, each of which would also be divided into pens.

Thus the heart of most exhibition breeders' establishments will be one or more fairly large shed(s) divided into several smaller pens, including some raised pens on the walls. The sizes of the pens will naturally vary according to the size of the breeds to be kept. Leading breeders will normally keep quite large numbers to maintain a strain for many years without excessive inbreeding problems. A surprisingly large flock of the smaller breeds of bantams can be kept in perfect show condition in the end part of a suburban garden with a little thought about layout.

Some exhibition poultry housing is very impressive, and very expensive. Books and magazines published in Canada, Germany, The Netherlands and the USA show luxurious hen houses that very few in Britain or elsewhere can equal. This is mainly a response to climate. The southern states of the USA, the southern counties of England and most of southern Europe – apart from mountainous regions – seldom have really severe winters to cope with. In much hotter climates the housing has to be designed to keep birds cool rather than keeping them warm: here, poultry housing mainly consists of a roof to keep off the rain and the heat of the sun, and the walls are mainly wire netting or weldmesh to allow a cooling breeze through.

Those on a tight budget, and those with limited DIY abilities, should still aspire to make their poultry houses look as tidy as possible. Some thought at the planning stage can give perfectly presentable results even when some of the materials used are 'recycled', and a coat of paint or coloured wood preservative can help a lot. Neighbours today are less tolerant than they were in previous decades, so do not create an obvious cause of complaints. If you have an existing building you can convert, this is a good start.

As this book deals with *exhibition* poultry keeping, it is fairly safe to assume that it is not being read by many complete beginners to poultry keeping. Virtually all readers will already be familiar with the main types of housing used by hobbyists. Larger sheds divided into pens will normally be home-made, or at least the internal pens will be. Smaller poultry houses with outside runs and movable 'arks' or 'fold units' may also be home-made, or may be bought. There are many designs advertised in the poultry and smallholding magazines every month.

Most exhibition breeders will use several types of housing, as all have their uses. However, some designs are better than others, so some suggestions of points to consider when choosing which to buy (or which to copy, for DIYers) are included below. Outside runs need suitable fencing to keep birds in the correct runs, and to keep predators out. Poultry keepers in Britain and western Europe have enough problems with badgers and foxes, so this chapter will not attempt to advise those keepers who live in countries with even larger and more powerful wild animals to cope with. In fact, some of the most difficult predators to exclude are the smaller animals in the *Mustelidae* family, such as mink, polecats, stoats and weasels.

THE IMPORTANCE OF A LARGE SHED

Even those with a very modest flock will have at least one moderate-sized shed at the heart of their set-up. As well as the stock housed within, there has to be somewhere to store poultry food, bales of wood shavings, carrying baskets and boxes, and assorted tools, spare feeders and so on. Rats and mice will soon sniff out and gnaw through food sacks, so the food must be stored in secure containers. Normal domestic dustbins (= US trash can) are cheap and are available at every hardware store. Those intending to bulk-buy wheat from local farmers will need larger containers, and purpose-made cornbins should be available from most agricultural supplies centres. Metal and plastic drums, such as those originally used for imported fruit juices, can be bought more cheaply from companies in the recycling industry. Do not buy any that have been used for dangerous chemicals. There is

no reason why all of this needs to be in the same building as any of the birds: in fact, your carrying boxes will keep cleaner if they are somewhere else, so the choice of storage space will depend on existing buildings available. There may be space in your garage?

Cage Construction
A bank of cages will be essential for show training, housing broody hens, hens with chicks and any other odd birds that need to be housed away from the main flock from time to time. The use of the word 'cage' should not be confused with battery cages. Battery cages are much smaller and have a wire floor, whereas the cages meant here have a solid wooden floor, and the bottom 6in (15cm) of the front is solid wood in order to retain a covering of wood shavings on the cage floor. It is therefore essentially like a small poultry house. The size of each cage can be varied according to the breed and the intended use.

When building cages like this, first remember that it must be reasonably easy to catch birds in them. Exhibition poultry should be as tame as possible, and the best way to do this is to handle your birds as often as possible. It is quite common for breeders to have three levels of these cages against a shed wall, where each level would be about 2ft (60cm) high, and 2ft (60cm) from front to back; this means that the floor of the top layer will be about 4ft (120cm) above the floor, and the access door will be at least 4in (10cm) higher, allowing for the litter-retaining board. Are you tall enough to reach birds in the top level? Especially if they do not want to be caught! Another consideration is the amount of light reaching the bottom layer of cages (or rather, the lack of it). This arrangement only really works in a fairly narrow shed, preferably with just a passageway in front of the cages, and with a lot of openable windows on the opposite side of the shed.

These cages can be of various lengths, say from 3ft (90cm) to 8ft (240cm), and their size can be made more flexible by having removable partitions. A 2 × 3ft (60 × 90cm) compartment will be ideal for sitting hens, hens with chicks, and show training. With a partition

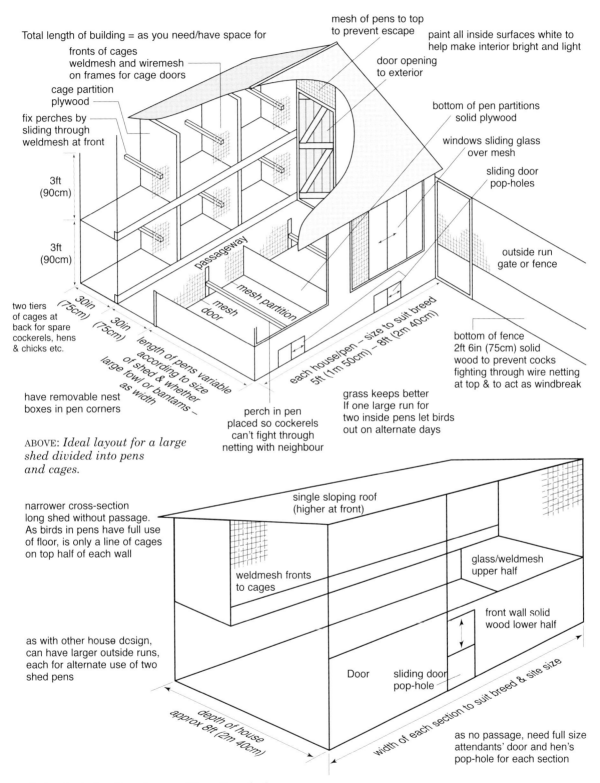

Total length of building = as you need/have space for

fronts of cages
weldmesh and wiremesh
on frames for cage doors

cage partition
plywood

fix perches by
sliding through
weldmesh at front

3ft
(90cm)

3ft
(90cm)

two tiers
of cages at
back for spare
cockerels, hens
& chicks etc.

30in
(75cm)

30in
(75cm)

length of pens variable
according to size
of shed & whether
large fowl or bantams –
as width

have removable nest
boxes in pen corners

passageway

mesh
door

mesh partition

perch in pen
placed so cockerels
can't fight through
netting with neighbour

mesh of pens to top
to prevent escape

door opening
to exterior

each house/pen – size to suit breed
5ft (1m 50cm) – 8ft (2m 40cm)

grass keeps better
If one large run for
two inside pens let birds
out on alternate days

paint all inside surfaces white to
help make interior bright and light

bottom of pen partitions
solid plywood

windows sliding glass
over mesh

sliding door
pop-holes

outside run
gate or fence

bottom of fence
2ft 6in (75cm) solid
wood to prevent cocks
fighting through wire netting
at top & to act as windbreak

ABOVE: *Ideal layout for a large
shed divided into pens
and cages.*

narrower cross-section
long shed without passage.
As birds in pens have full use
of floor, is only a line of cages
on top half of each wall

weldmesh fronts
to cages

as with other house design,
can have larger outside runs,
each for alternate use of two
shed pens

single sloping roof
(higher at front)

glass/weldmesh
upper half

front wall solid
wood lower half

Door

sliding door
pop-hole

depth of house
approx 8ft (2m 40cm)

width of each section to suit breed & site size

as no passage, need full size
attendants' door and hen's
pop-hole for each section

A simple design for a long multi-section shed.

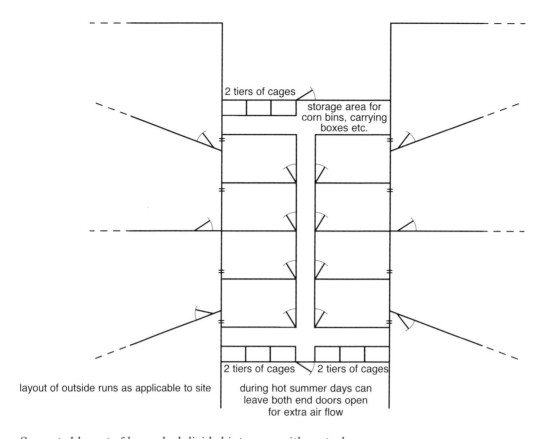

2 tiers of cages

storage area for corn bins, carrying boxes etc.

2 tiers of cages

2 tiers of cages

layout of outside runs as applicable to site

during hot summer days can leave both end doors open for extra air flow

Suggested layout of large shed divided into pens with central passageway.

removed, making a 6ft (1.8m) long pen, you have space for growers that are not quite ready for the great outdoors, or for winter housing of a breeding group of bantams. If large fowl are to be housed, especially exhibition cockerels, more headroom will be needed. Only two layers will be possible, each at least 2ft 6in (75cm) high and deep. The minimum cage length would be 4ft (120cm), and this is too small for cocks intended for shows.

The best front for this sort of cage is fine mesh on a wooden frame, because this will then keep in day-old chicks. Those pens that are not likely to be used for chicks can have a differently constructed front. Remember to have doors that are large enough to allow you to catch birds easily, and for cleaning out.

There is little point in being more precise in describing these 'penning rooms' as each will

be different, according to the size, shape and materials of the shed or other previously existing building used for the purpose. It is a good idea to visit some established breeders before you start your own construction to copy the best ideas you see, so as to avoid obvious problems.

SEPARATE SUMMER AND WINTER HOUSING

It is not a matter of chance that the most elaborate housing is seen in countries with the coldest winters: a solidly constructed building with electric lighting and heating is a necessity, not a luxury, in these conditions. Fortunately most breeders have fewer birds to house during winter, as only the best will be retained for breeding and showing. In the early spring, when hatching has started, the chicks will be inside, under heat lamps anyway. By the time these chicks are old enough to benefit from access to outside runs, the weather will be warm enough for wooden huts and arks to be used, as in warmer climates. Breeders with enough space may also have arks to move some of their adult stock out for the summer.

Although separate winter housing is not common in Britain, it should certainly be considered, if possible, by those living in exposed, windy situations. By late summer the chicks will be nearly mature, and there will still be the best of the stock bred in previous years. This will add up to far more than can be accommodated in the winter housing. Some will have to be sold or culled, a useful discipline for those who find it difficult to make tough decisions. The winter housing can still be used through the summer, of course. If the site layout permits, some or all of the pens can have 'pop holes' (for those of you who are novices, a pop hole is a small, chicken-size hole with a door) allowing access to outside runs for use during summer, and those times in the winter when the weather is not too bad.

Planning Permission

If a substantial building is to be built, it is necessary to check with your local authority planning department. The rules vary in different places, being much stricter in conservation zones or areas of outstanding natural beauty, and there will be different rules in each country. Thus it may be permitted in some areas to erect free-standing sectional sheds, but not to build a permanent (with foundations) building the same size. It may also be permitted to have two small sheds, but not one large one. A thorough knowledge of the planning laws in your area can prevent a lot of problems, and even legal expenses later.

Purchasing a Building

Second-hand sectional buildings can often be bought a lot more cheaply than new ones; check the advertisements in local newspapers,

Exchange & Mart, *Free-Ads* papers, *Yellow Pages* and so on for possible suppliers. Suitable sheds may not, however, be as easily found in the early twenty-first century as they were back in the 1970s. Modern 'Portakabin'-type units do not blend into farms or gardens as easily as traditional wooden 'building-site' huts did. Many farmers and hobbyists in previous decades were also able to buy ex-Ministry of Defence sectional wooden barrack sheds, though the supply of these is probably exhausted by now.

HOUSES AND RUNS IN A LIMITED SPACE

Very few hobbyist poultry keepers have as much space as they would like, and even those with quite large country properties will prefer not to have large areas reduced to bare earth or mud for much of the year by the constant scratching of hens. In areas where the winters are not too cold – in south-west England, for example – most breeds will be perfectly happy for much of the time in a suitable-sized shed with the upper half of the front made of strong weldmesh. They can be let out as much as the weather and the grass in the run allows. Two or three such pens can share the use of one larger outside run. The birds will be better having two days a week out on grass than seven days a week out on bare earth, the inevitable fate of small outside runs. This arrangement is normal in suburban gardens where the hens have to share the lawn with the rest of the family.

In colder climates, and where more delicate breeds are to be kept, the same principle can be used as far as sharing the grass run is concerned, but the housing must be different. Breeds with a large comb, such as Leghorns, can really suffer in long, cold winter nights; they can lose the tip of their comb from frostbite. The solution is to divide their accommodation into two parts: a small, closed-in house for roosting and the nestboxes; and a covered aviary-type run where they can scratch about during the day. In a confined space their own bodyheat will

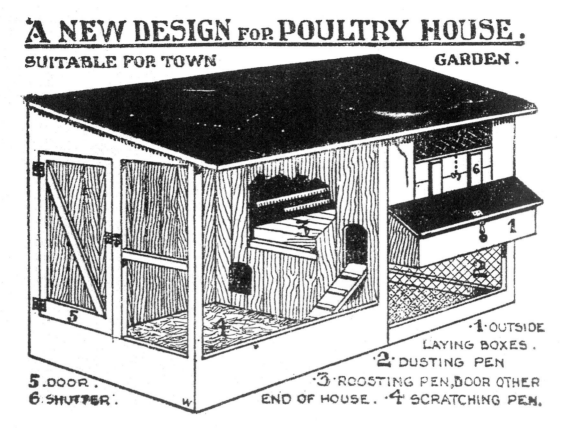

A NEW DESIGN FOR POULTRY HOUSE.
SUITABLE FOR TOWN   GARDEN.

1 OUTSIDE LAYING BOXES.
2 DUSTING PEN
3 ROOSTING PEN, DOOR OTHER END OF HOUSE. 4 SCRATCHING PEN.
5 DOOR.
6 SHUTTER.

'A New Design for a Poultry House', circa 1919, and still a good design. From Fortunes from Eggs, *published by The Karswood Company, manufacturers of 'Karswood Poultry Spice'.*

raise the house temperature enough to prevent comb problems in all but the worst freeze-up. Besides, if very extreme weather is forecast, many breeders will arrange emergency measures: this may even include a few large combed cockerels spending the night (in baskets or show cages) in the kitchen!

For a breeding group of, say, a cock and six hens of large Anconas, Leghorns or Spanish, the roosting house might be a converted 7 × 5ft (approx 2 × 1.5m) garden shed, and the aviary run either 7 × 10ft (2 × 3m) or 5 × 15ft (1.5 × 4.5m), depending on the general layout. This run should have a corrugated plastic sheeting roof, and the sides should be mostly wire netting apart from the bottom 30in (75cm), which should be solid to prevent cock birds from fighting through the wire with others that are running outside.

CONVERTING A GARDEN SHED TO A HEN HOUSE

Purpose-built poultry houses are usually a lot more expensive than similar-sized garden sheds. This is partly because garden-shed manufacturers are much larger operations than specialized poultry-house companies, and partly because poultry houses have fixtures and fittings such as perches and nest boxes. These will have to be made by the buyer, which may be a positive advantage as built-in nest boxes are often too small for the largest pure breeds, or for breeds with long tails. Removable nests are useful, as broody hens can be taken away for incubation duties in the nest in which they are already happily settled. Built-in nest boxes are often favourite places for blood-sucking red mite infestations, and even with sprays and strong disinfectants, these can be difficult to eradicate. A washable plastic drum or disposable strong cardboard box in a dark corner, with a couple of bricks in front as a doorstep, may look rather improvised, but has practical advantages.

The perches fitted in manufactured poultry houses are usually fine for bantams and some smaller large breeds, but are too narrow for the heaviest large breeds: blistered and dented breasts can be the result of too much body-weight concentrated on a narrow perch. Also, the ideal height of a perch is different for certain breeds, and some varieties are even better sleeping on the floor. (Perches are discussed in detail later in this chapter.)

A pop hole will also have to be fitted when converting a garden shed, a job that is not too difficult, even for those with limited DIY abilities. It can, of course, be installed in whichever side suits your overall layout, which may not be the case in a bought hen house. Although you may be intending to keep one of the smaller breeds at present, you may change to a larger variety in the future, so make the pop hole large enough for any possible breed. The width would usually be the gap between the uprights in the shed frame, normally about 15 to 18in (40cm). Sliding doors, horizontal or vertical, are normal for pop holes. If the pop hole is dividing a house from a covered run, it is best to have a vertical sliding door operated by a string and hooks: it can then be opened and closed from outside, saving valuable seconds every day if you are in a rush to go to work. A more secure door will be prudent if the pop hole is to access an outside run where foxes or badgers might be about overnight.

Finally, garden sheds do not normally have openable windows, or even enough windows for use as a poultry house. Existing windows can simply be covered with wire netting to prevent birds trying to fly through the glass (chickens can be very, very stupid!), and a part

of some wall sections removed, and the opened parts also covered in wire netting, to provide fresh air and more light.

The house can be made much lighter if the inside walls are painted white. Internal painting gives the added benefit of sealing cracks where mites could hide, and gives a washable surface for extra hygiene. Leading exhibitors of white-plumaged varieties are especially likely to aim for the cleanest possible conditions.

If, after a few years, it is decided to give up keeping poultry, a converted garden shed is easily adapted to some alternative use, whereas most bought poultry houses are too low for anything else. Besides, a poultry house that is high enough to stand up in normally is much more comfortable for cleaning out and for other poultry-keeping duties.

BUYING TRADITIONAL-TYPE POULTRY HOUSES

There are many designs of traditional poultry houses advertised in poultry and smallholding magazines, and some can be closely examined at agricultural shows and farm parks, agricultural merchants and elsewhere where they are on sale. Most of these manufacturers make a range of sizes and designs, both of houses intended to be static, and movable arks/coops/folds (discussed below).

Static Houses
Static houses are normally used with an outside grassed run. The following are a few design features that should be considered:

- Plywood-constructed houses are often cheaper than those made of feather-edged or tongued-and-grooved boards. Plywood may not look as attractive, but it is much more practical because it is less draughty, stronger, and it provides fewer hiding places for red mite.
- Is the main door large enough for you to catch birds and clean the house easily?
- Are both the main door and the pop hole on the side that suits your intended layout? If

not, is the manufacturer willing to make a modified design to order? Some smaller companies may do this, especially if you are going to order several similar houses, perhaps four 'left-handed' and four 'right-handed' houses to be placed with their runs on both sides of a central pathway.

- Is the house large enough for the birds to be kept inside when the weather is very bad? Be aware that a house that is claimed to be for 'twelve hens' will probably hold far fewer exhibition birds.
- Are there enough windows, covered in glass and/or wire netting, to provide adequate light and ventilation? This is probably the main deficiency in the designs on sale at the time of writing (2004).
- New materials: also in 2004, a few poultry houses appeared made of plastic and materials other than the traditional wood, plywood and several types of roof sheeting. There will probably be more of these in the future. When new they will probably be as good as a similar sized, but traditionally constructed house, and no doubt a lot easier to keep clean. However, plastic houses may be difficult to repair after inevitable breakages.

Movable Arks

These are also called 'fold units', and consist of a combined house and run unit that should be moved every day, or at least every other day. They have traditionally been made in triangular cross-section, allowing for simple and economic construction. When used on a farm or smallholding, being moved around an orchard or paddock, the steeply sloping sides/roof prevent sheep or goats from sitting on top, which they often do on flat-topped houses or runs. Such triangular arks are not so good for exhibition birds with large combs and tails because the limited space at chickens' head height will almost certainly break tail feathers and cut or scratch combs. This will not be a problem for breeders of Wyandotte bantams and other compactly built breeds.

Fortunately there are some designs with vertical sides and a less sloping roof for those with large-tailed poultry, and no goats. Some designs have a separate house and run section, which makes the larger and heavier models easier to move; also some designs have wheels, which will help on firm ground at least.

The run part of most arks is simply constructed of wire netting over a wooden frame. This is fine in sheltered situations, but will need the addition of some plastic sheeting on windy sites. A few designs have part-plywood-clad runs, which are excellent.

One UK manufacturer produces (in 2004) a design of ark in which the house part is a tunnel along the top of the structure. These are not suitable for exhibition stock because the house part, the tunnel, is too cramped. A few other companies produce movable aviary type arks, in which the house part is considerably larger, and the run is almost high enough for people to stand upright in. These are expensive and excellent – but be aware that, because they are so much higher, they may get blown over by strong winds in exposed situations.

Arks do require the extra work of being moved every day or so, but have the advantage of securely containing those breeds that are inclined to fly over fencing, and of protecting stock from foxes. They also enable complete control of grass wear, they require less material than the permanent fencing of open runs, and they look much neater in a garden than do most open runs.

With their fairly small house sections and mostly open mesh runs, and in view of the muddy state to which grass can degenerate over winter, the best way to use arks in the winter months is to keep them on a concreted area, or under an open-fronted farm building if one is available.

RUN FENCING

Outer perimeter fencing must be adequate enough to keep out foxes, badgers, and any other potentially harmful creature, as well as keeping your poultry in. Fencing within the establishment, to divide one group from the next, should be solid up to cockerels' head

'Broodie Ark', made by Fisher's Woodcraft, Doncaster. Similar designs in a range of sizes are made by several manufacturers, though some of them, especially smaller models, will not be suitable for potential show cockerels with large tails. For a hen with chicks or a group of growers, this model is ideal.

'Garden Crop house with run', made by Fisher's Woodcraft, Doncaster. A very useful house and run that can be moved every day or two around a paddock or large lawn. The run could be improved by the addition of some plastic sheeting as protection against wind or rain.

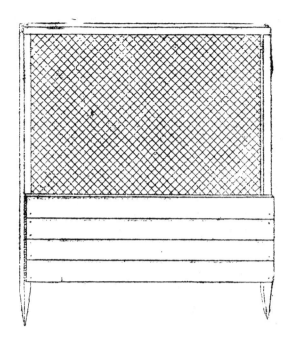

Standard run fencing, in this case in the form of a movable hurdle for temporary runs, from Poultry Appliances, and how to make them *by L.B. Collier, circa 1920.*

height, to prevent them fighting through the wire netting. A solid lower part is also useful for outer fencing as well, partly as a wind-break, and partly because the old adage 'to see is to want' applies to foxes as well as people. Predator animals will often attempt to dig their way in, so some strong netting or other barrier needs to be buried beneath the perimeter fence. At the top of the fence there should be angled arms to hold several strands of barbed wire to prevent the more athletic local foxes from entering. All run fencing should be at least 6ft (1.8m) high, not counting the barbed wire around the perimeter.

The ideal grass-run area is 6sq yd (5sq m) per large fowl, or about half that per bantam – though very few breeders, in the UK anyway, have enough land to provide anywhere near that much: this is equivalent to a run 16.5 × 33ft (5 × 10m) for one large fowl cock with nine hens. This is why it is recommended to have one grass run for two or three houses, and

allow each group out two or three days a week, when the weather permits.

Outside Runs for Game Fowl Breeds
These are traditionally called 'fly pens', and typically are covered with wire netting over the top. This is because Game cocks, especially Oxford-type Old English Game, are athletic enough to fly over even 6ft (2m)-high fences. These birds need to be physically fit for exhibition, as a major requirement in a show class is muscular tension apparent when the birds are handled. Many breeders install swing perches in these aviary-type runs. Rigid wire is probably better for hanging these swings than chain, but chain or rope is acceptable if the swing perch is heavy, a stout round fence post perhaps. The birds will exercise their leg and breast muscles every time they fly up to the perch, and they have to keep balance when it moves. If there are several such runs in a row, with solid partitions up to 3ft (90cm), the cocks will repeatedly fly up on these swings to threaten the other cocks they can see. It is important (and obvious) that these swings are well away from the sides of the runs so that the cocks cannot actually peck each other through the wire netting.

Shade for Outside Runs
Continual exposure to bright sunlight causes white plumage to go a yellowish colour ('brassy'), and buff and blue plumage to fade. Shade can be provided by shrubbery or horticultural shade netting. Your local (plant) nursery will have details of suppliers of this netting.

PERCHES

One size certainly does not fit all breeds, and most perches fitted in commercially made poultry houses are too narrow. Perches just 1.5in (4cm) wide are only suitable for bantams, and the larger breeds of bantams, as the smaller breeds of large fowl, need 2in (5cm) perches. Heavy breeds of large fowl need 2.75in (7cm) perches to give more area for the bodyweight to rest on. All perches should have rounded edges,

'Rothwell' house and run unit, made by Fisher's Woodcraft, Doncaster. The covered run makes this an excellent design for blue, buff or white-plumaged varieties that need shade.

and it is better if they fit into brackets rather than be nailed in place: this is so they can be easily checked for red mite infestation, or can be completely removed when the pen is used for rearing young birds. Narrow perches, and any perches for young-stock will result in a bent keel bone and breast blisters. Japanese (Chabo) bantams are not able to perch because of their extremely short legs, and many specialist breeders of heavily feather-legged breeds prefer not to provide perches. Chickens roosting on a perch will naturally huddle together and by definition will all be in a straight line, and they are almost certain to break each other's foot feathers in the process. Such damage is much less likely when they are sitting at various angles to each other on the floor.

The average height of perches is about 30in (75cm) above the floor, high enough for the birds to walk under them without hitting their heads, and so avoiding possible damage to combs or crests. Anyone keeping a large group together and thus requiring more than one perch in a house, should have all perches at the same height because hens may fight for a place on the top level. Yokohamas, Sumatras and other long-tailed breeds need higher perches, high enough for the tails not to reach the floor. Again to prevent tail damage, all perches should be well away from the walls of the pens; the required safe distance will vary according to the breed. Indian Game, Ixworths and other very heavy breeds need much lower perches because they can easily injure their legs and feet when jumping down from high perches. Bumble foot, where the soles of the feet are swollen, is very common in heavy breeds.

NEST BOXES

Most commercially made hen houses have nest boxes that are fine for hybrid layers, but are too small for exhibition large fowl breeding stock. Ideally nest boxes should be large enough for cock birds to get in, which they often do. This is normal behaviour for wild jungle fowl, and they do so to check the nest for hidden snakes or other predators before the hen enters to lay. No one has yet found a way of explaining to a domestic cockerel that the nest box in your hen house will not have a venomous snake in it, or that he is not a wild jungle fowl.

Hens also still imagine they are wild jungle fowl, and given a choice, will favour the darkest nest boxes in the corners of the house. If individual nest boxes are made, they will be ideal if part of the front is solid so the hen can feel safely hidden. Removable individual nest boxes are excellent where it is intended to encourage hens to go broody to hatch some eggs. This highlights one great difference between breeding flocks and laying flocks, in that breeding hens are encouraged to go broody, whereas laying hens are discouraged from doing so.

'Marsden' house, made by Fisher's Woodcraft, Doncaster. A solid traditional design that, with minor variations, has been successfully used by thousands of poultry keepers for over a century. Several manufacturers make similar houses in a range of sizes. Breeders should note that the size suitable for twenty hybrid layers would only house half that number, or less, of large exhibition Sussex or other heavy breeds.

FEEDERS AND DRINKERS

There are several basic types of feeder and drinker on the market – though many breeders still prefer to use dog bowls or other containers, perhaps shallow bowls found at a car boot sale. In fact these alternatives can be much more suitable than 'proper' feeders and drinkers for breeds with large combs, wattles, crests or beards.

Many poultry books give an allowance of the trough space needed per bird, or the number of drinkers needed per 100 hens. It is equally important that there should be enough space for all birds to feed simultaneously, although with exhibition birds normally being kept in small groups it is seldom a problem. Pure breeds are often more aggressive than commercial hybrids, but an extra few centimetres of trough space will not make much difference. If it is not possible to remove either the bully or the victim to another group, the best solution is to scatter a handful of food around the floor to distract some of the birds away from the trough. Bullied hens can often be trained to feed away from the others, in a nest box perhaps.

All types of poultry, as any livestock, should have access to clean drinking water at all times. Apart from the obvious welfare issues, it should be remembered that eggs are mostly composed of water; so no water, no eggs. Like ourselves, poultry drink more in hot weather, so it may be necessary to provide extra, or larger, water containers in the summer.

Tube Feeders

These consist of a cylindrical storage tube with a trough around the bottom. There are many different types available, some galvanized steel, some plastic, some for inside use, others with rainguards for outside use. Because adult show birds are usually kept in small groups, these feeders are only suitable for larger groups of growers. The reserve of food will attract rats and mice if feeders are left with food in every night, and it is much better to just give enough for each day, and collect uneaten food every night. Note that outdoor feeders with rain covers do not have enough headroom for large combed cocks.

Fount Drinkers

Fount drinkers rely on a partial vacuum to hold the water in; they have been in use for a very long time. In Victorian times they were made of pottery and had a small cup-shaped bowl at one side of the base; some are quite ornate and have become valuable collectors' items. For many years they were made of galvanized steel, 'Eltex' being the main manufacturer in the UK. These are still available and are very good quality products, but are rather expensive compared to the current plastic versions. There are also drinkers of this type, again made by 'Eltex', with metal bases and glass bottles to hold the water. All types are made in a range of sizes from 2 pints to 3gal (1–15ltr). Most fanciers will have a selection of sizes to suit every class of stock from chicks to the largest adults. When used for half-grown or adult birds it is best to stand drinkers up on some bricks to keep the water clean.

Fount drinkers are less good in the depths of winter as they can freeze solid, which can split plastic drinkers. Also, the plastic can become very brittle after a few seasons exposed to extremes of heat and cold.

Automatic Water Systems

There are several systems available, the most cost effective using plastic dome-shaped drinkers that are suspended from the house ceiling. Water is fed down in pipes from a header tank to a valve at the top of each drinker; the valves are spring-operated, releasing more water as the hens drink. The disadvantage to any automated system is that it is not reliable in freezing conditions, so these systems are only suitable for large buildings divided into several pens. The main operating problem with these drinker systems is when valves stick open, which can even lead to flooded hen houses. Valves should be regularly checked, filters fitted to the outlet pipes of the header tanks, and the header tanks should be covered to prevent dirt entering the system.

Open Troughs for Food or Water

Large dog bowls, pottery or plastic and available at every pet shop, are ideal for small groups of exhibition birds, especially those with large combs. Leghorn or Minorca cocks can hit or rub their combs on narrow troughs, and those with 'anti flick' grids on top are also not suitable for adult exhibition birds, although they are fine for young-stock. If dog bowls are used, choose a type that is least likely to be tipped over by the birds, and buy the largest size available. Many troughs have a lip on the inside top edge to minimize food wastage.

Plastic can get brittle and split with age, steel can go rusty, and pottery can chip. In all cases, discard any troughs that have jagged edges, which might catch and injure the birds.

Drinkers for Crested Breeds

Polands and other breeds with large feathered crests need special drinking arrangements to avoid crushed or soaking wet crests. The ideal, especially for individually caged show birds or pedigree mated pairs, are budgerigar drinkers in the largest size available, because the birds can dip their beaks in the small trough part at the bottom without getting their crest or beard wet. Larger groups can have their water in open bowls with netting over the top: they dip their beak through the mesh to drink and the mesh holds the crest out of the water. The bowl/trough capacity should be very generous as there must be enough water left at the end of the day for the birds still to be able to reach it.

Feeders and Drinkers for Tall Breeds

German Langshans, Malays, Modern Game, Modern Langshans and Shamo are required to pose standing well up at shows. They can be encouraged to do so by having high food and water containers, even more so if they are given extra tasty treats in these high pots. Then when they are at a show, it is to be hoped that they will stand up when the judge approaches in the hope that he will be giving them something as well.

BREEDS WITH SPECIAL NEEDS

Housing and Fittings for Japanese (Chabo) Bantams

As said above, Japanese bantams cannot manage perches, and because they normally sleep on the floor, their houses should be cleaned out as frequently as possible; this is especially true with Whites and Black-Tailed Whites. Conventional nest boxes are not suitable for Japanese hens because they might have trouble getting in them, and when they do they will probably damage their tail feathers on the inside top of the box. These hens normally just lay in the darkest corner of their house or cage. It might be worth fitting a privacy screen, although there is a risk that the Japanese might try to perch on it.

Housing and Fittings for Feather-Legged Breeds

Much the same considerations apply as for the Japanese, with the added requirement of ensuring there are no places for the foot feathering to get rubbed and broken.

Needs of Breeds with White Ear Lobes

White lobes, especially on Rosecomb Bantams and Minorcas, are prone to becoming reddened or blemished by brown scabs. The large lobes on these two breeds seem to be worse than other breeds, possibly because the relatively hard tissue they appear to be composed of has a poor blood supply. White-Faced Spanish have a much more extensive area of white skin, but any scabs on Spanish seem to heal up quickly: their faces are softer, and they are better supplied with blood, and hence healing white blood cells.

Dirt and grit are the probable causes of lobe blemishes, so these breeds should be kept as clean as possible. The reddening of the lobes is a response to sun and wind; therefore outside runs should ideally be sheltered by high fences or shrubbery.

CHAPTER FIVE

Nutrition

Exhibition poultry are mainly fed on crumbs, pellets or meals designed for hybrid layers, broilers, turkeys or ducks, and for pheasants reared for shoots. Pheasant pellets are similar to turkey rations. Feed such as mixed corn is fed as well, but if too much will lead to nutritional deficiencies.

There are obviously differences in growth rates and egg production between exhibition birds and commercial stock, but the many decades of intensive research into poultry nutrition, much of which has been either financed by, or directly done by the feed companies, means that commercial rations are much more accurately suited to their nutritional needs than any mixture an amateur could produce. In addition to giving direct instructions on the feeding of show poultry, this chapter should enable fanciers to think logically about using different diets to achieve specific results.

GENERAL PRINCIPLES OF COMMERCIAL RATION FORMULATION

Three sets of basic information are combined to arrive at each ration formulation: the nutritional requirements of the relevant type of livestock; the nutritional content of the available ingredients; and the cost of the raw materials.

Nutritional Requirements of the Relevant Type of Livestock

The type of livestock might include:

- hybrid layer-type chicks, 0–8 weeks;
- hybrid layer-type growers, 8–18 weeks;
- hybrid laying hens, small-bodied white egg type;
- hybrid laying hens, medium-bodied brown egg type.

This list could be continued for breeding stock, broilers, turkeys and ducks of various ages. Brown and white egg layers have significantly different nutritional requirements because white-egg birds lay as many eggs as brown eggers, perhaps more, but have a lower body-weight and eat less food.

There have been many hundreds of test flocks reared over the years to arrive at the present formulations, but even so, there are differing opinions on many details. Broiler chicken companies typically produce many millions of birds each year (that is, each company: it's *billions* of broilers worldwide). Very small changes in growth rates have major economic results: thus, integrated broiler chicken companies do the whole process from breeding stock, via incubation and rearing to processing, and each batch has to be at the correct weight range on a specific day, or the whole system falls apart. The customers, supermarkets or fast food chain, need specific sizes of birds, the sheds have to be cleared (ready for the next lot from the hatchery), and the processing factory is probably working twenty-four hours a day, seven days a week. Chicken feed is the main cost of these operations. Broiler breeding stock nutrition is also closely controlled.

Nutritional Content of the Available Ingredients

Having discovered the percentages of protein, fat and so on, and the vitamin and mineral levels and suchlike required for a ration, the same set of information has to be discovered for wheat, maize, soya meal and the rest of the ingredients. General tables of bird requirements and raw material contents have been available for many decades, but individual batches are tested as levels can vary significantly according to the exact cereal variety and the growing conditions at the time. In this era of global warming there have been some serious droughts that have affected crop quality.

Some useful ingredients are by-products of vegetable oil production or other industries. The BSE crisis has stopped the use of meat meal, meat and bone meal and other products recycled from the meat industry. But to be fair it was used at least back to the 1930s, perfectly safely as far as anyone knew. The use of a modest percentage of meat meal in farm livestock diets to achieve the required protein level was not invented by modern 'factory farmers'. Protein is now mainly obtained from soya bean meal. Some synthetic amino acids, vitamins and so on are needed because they only occur naturally (at adequate levels) in animal protein.

It has been discovered, sometimes as a result of unfortunate experiences, that some raw materials contain substances that are toxic, or which greatly unbalance a meal if too much is used. They are perfectly safe, and often good value for money, as long as they only form 5 per cent (or whatever is recommended) of the ration. Be reassured that the testing requirements are very strict now, and no feed companies are going to take any chances.

The Costs of Raw Materials

One of the reasons that bags of poultry feed do not list ingredients in any great detail is that the manufacturers would have to change the labels almost every day if they did. It is now simple to run a computer program to calculate the cheapest way of achieving each formulation as market prices rise and fall, keeping to all pre-set limitations. Changes in prices and formulations are usually a matter of fine tuning, but the smallest of changes can make a huge difference to the finances of a company producing hundreds of tonnes a month.

PURE BREEDS ARE DIFFERENT

Commercial hybrids, both broilers and layers, develop and produce in their different ways much more quickly than pure breeds. This is partly genetic, a result of the intense selective breeding used to create them, and partly a natural consequence of the different housing and management methods. Hybrid layers start producing at about twenty weeks of age, whereas most pure breeds will be not much more than half grown at this age. It should therefore be obvious that pure breed stock should not be changed from grower to layer ration until they look as though they are approaching maturity. This will be different for each breed, sometimes varying quite a lot for different groups of one strain. Commercial stock is normally kept in controlled environment housing, especially regarding day length, and is genetically very uniform anyway. Hobbyists' birds are naturally housed, and so react to varying temperatures and daylight. Many inbred strains will also be genetically stable, but on the other hand most hobbyists will have more than one breed, and will be rearing mixed groups.

Although exhibition stock will not be laying or growing as much as hybrids, they are often less efficient at absorbing nutrients. The very tiny bantam breeds eat very little, especially in the first month, so these chicks still need the best possible start in life. At the other end of the weight range, large Orpingtons, Jersey Giants and others need to be as big as possible if they are to win at the shows. Therefore many exhibition breeders use higher protein broiler starter, or even very high protein pheasant or turkey starter crumbs. These rations may be too much for use over the whole 'chick period', but it may be a good idea to buy a few bags of pheasant starter for the first two weeks, then change to normal chick crumbs for another eight weeks, and finally change to growers pellets until the pullets start maturing. The change from growers to layers rations can even be left until the first eggs are laid in a group of pullets; they can 'borrow' calcium from their bones to make a few eggshells, and any depletion can quickly be replaced from the limestone flour content in layers meal/pellets as soon as they are started on it.

FEEDING ADULT MALE BIRDS

The following hypothesis is obvious really, but very few poultry breeders ever think of it for themselves: layers pellets contain 3 per cent limestone flour to make egg shells. But cock birds do not lay eggs, and if fed entirely on layers meal or pellets (designed for the hens), the excess calcium may upset their metabolism since the levels of calcium, phosphorus and vitamins A and D are related, and an excess or shortage of one will affect the others. It is not generally possible to give cocks a different diet while they are in with their hens, although as the males of some breeds are significantly taller than the corresponding females, special high hook-on food pots can be tried to give the cocks an extra tasty (low calcium) treat.

Many, probably most, exhibition and hobbyist poultry keepers feed mixed corn as well as pellets, and it is quite possible that cock birds will instinctively eat a higher proportion of corn than the hens. This may create another problem, however, in that the cocks may not get enough protein, vitamins or other essential nutrients.

Most hobbyist breeders keep several males of each variety and fertility is improved by swapping cocks at least once a week. This is partly a behavioural response to having a week away from 'his' hens, and partly a result of having a general nutritional recovery. Many cocks do not eat enough when they are with a large group of females, but if they can be given a lower calcium diet, probably based on growers pellets, during their rest and recovery periods, these can be even more beneficial. Where cocks are kept with only one or two hens, however, it is probably better to keep them with their mates all the time as they usually become very attached, and the negative effects of the stress caused by being parted from them will be worse than the nutritional benefits of some 'time off'.

FOOD WASTAGE

Food is the main cost of poultry keeping, so you do not want your birds to throw it all over the place. If they do so, it may be because they do not like the food available (*see* Food Palatability below), or the food troughs are poorly designed. Rats and mice will soon sniff out a free lunch, and they can bring diseases with them. To minimize this, get into the habit of going round with a bucket to collect up all uneaten food from the troughs when you shut your birds in at night. Some factors that affect food consumption include: the palatability of the food; the temperature of the environment at the time; feather cover; social stress; disease; and any change in diet.

Food Palatability

If chickens do not like the food they are offered, they will only eat as much as they need in order to avoid the greater discomfort from hunger. Thus they will reduce consumption enough to reduce growth rate or egg production – and remember that potential growth which does not happen when birds are young cannot be corrected later: such birds will be permanently stunted. Wastage can increase if they are given a mixture in which they like some parts (cut maize?) more than others (barley?): they will expertly eat their favourite ingredients, and throw the rest out of the trough.

Taste and smell are poorly developed senses in poultry, which mainly rely on sight and touch, or 'mouth feel'. One reason that maize is a favourite is its bright orange colour. Wheat is more popular than barley, and whole oats are very unpopular, because of the grain shape (oats are very pointed). Hens do also have enough sense of taste to refuse stale or musty food, and long enough memories to remember not to eat anything that gave them a stomach upset.

Wet mash (just damp, not sloppy) is most popular, followed by pellets, then coarse meal, and finally finely ground meal. Commercial layers, whether in battery cages or 'barns', are fed dry meal to keep them occupied. It takes them all day to eat enough, so minimizing fighting or even cannibalism in the flock. Most hobbyist poultry keepers feed pellets, the best choice for free range and small groups of exhibition birds housed inside. Wetting pellets to make them more attractive is not normally done, but is a useful ploy if your birds are reluctant to eat them, when they are ill for example. Wet mash, including dampened pellets, soon starts to go off, especially on hot summer days. If this is fed, only give enough for a day, at most.

Food Consumption and Environmental Temperature

All types of poultry react to environmental temperature in exactly the same way as we do; so, for example, they eat more when they feel cold: we normally change the content of our diet, as well as the amount, according to the season – protein, vitamin and mineral requirements remain constant, but energy needs (mostly carbohydrates, the remainder fats) rise as temperature falls. The solution for most hobbyists is to feed a higher proportion of pellets (and less mixed corn) in summer than in winter. In very hot conditions food consumption can be depressed enough to reduce growth rate or egg production, as well as stressing the birds. In these circumstances, try to provide an extra high protein diet, and take all available measures to provide shade and ventilation; in extremely hot conditions when housed inside, the ammonia gas released from deep litter makes the situation worse, and the lung irritation can lower resistance to respiratory diseases.

Food Consumption and Feather Cover

This is related to the environmental temperature changes described above, and is more likely to be a consideration with commercial layers, often nearly bald, than exhibition birds. These, we all hope, will have every feather perfect. However, all birds moult, and some autumn nights can be very cold, and breeders of Transylvanian Naked Necks might find a higher energy diet better for most of the year.

Food Consumption, Social Stress and Trough Space

The smallest and weakest of a group of birds may not be getting enough to eat if there is not enough trough space, as they will be driven off by the stronger birds. To be extra sure that all are well, throw a handful or two of food on the floor, well away from the food trough. Lack of trough space is seldom a problem with adult exhibition birds because they are usually kept in small groups, but larger groups of growing stock may need watching, especially if there are breeds of very different sizes and personality types – Game breeds with Cochins, perhaps.

Food Consumption and Disease

Sick birds will usually eat less than normal, which makes them even less able to recover. If the problem is intestinal worms, food consumption will increase, but body condition will worsen. In addition to seeking veterinary treatment, change to the most palatable food available.

Change of Diet

All changes of ration, especially from chick crumbs to growers pellets, should be gradual. Mix the two rations together, in gradually changing proportions, over at least a week. Poultry sometimes even notice, and are reluctant to eat, a change from one brand of pellets to another.

WATER

All poultry must have clean water available at all times. Water containers must be washed regularly. The comments above about food trough space also apply to water containers. Polish and other crested breeds need water containers that will allow them to dip their beaks in the water, but keep their crests out. Old water containers with jagged edges, or metal that has gone rusty, or plastic gone brittle, must be disposed of and replaced with new, smooth-edged ones. It is better to do this before one of your potential champions tears a wattle or loses half its beard.

It is important to be aware of the factors affecting water consumption because your normally adequate water containers may run out, stressing your stock in some circumstances. These factors include: environmental temperature; egg production; type of food; and drugs in the water.

Environmental Temperature: The ambient temperature is the most obvious factor to affect water consumption. Like you, birds drink a lot more in hot weather, so where there is any risk of water running out when conditions are hot and dry, provide additional drinkers.

Egg Production: Eggs are mainly composed of water, so hens will drink more when they are laying.

Type of Food: At the two extremes, birds fed on a diet including wet mash and access to grassy free range will drink less than a flock fed on dry meal in a deep-litter house. The actual water intake is the same in both cases, it just seems less in the first flock because they are getting part of their water from the wet mash and the grass. Some changes in rations can affect water consumption; this is a biochemical reaction to different protein, fat or mineral levels.

Drugs in the Water: If there is a disease problem, and a drug is supplied by your vet which has to be given in the drinking water, the colour and taste of the treated water may depress water consumption.

GRIT

There are two types of grit, soluble and insoluble, and they have different functions.

Soluble Grit: This dissolves in the gut to provide calcium for eggshells. As layers meal/pellets contain limestone flour, it is not normally necessary to supply soluble grit. Many exhibition birds are not very good layers, so the calcium in the diet will probably be enough. If you have a problem with soft- or porous-shelled eggs, it might be worth trying some soluble grit. The main type is crushed oyster-shell grit, with cockle-shell, mussel-shell and clam-shell grit being alternatives. This may not work, the eggshell problems may have a different cause, but sometimes the other substances in sea shells help the problem.

Insoluble Grit: Flint or granite grit collects in birds' gizzards to grind food, especially whole grain, before digestion. Free-range stock will usually find their own small stones when pecking about. Both types of grit can either be provided in separate pots, or occasionally sprinkled over the food.

SPECIAL DIETS FOR EXHIBITION BREEDS

Breeds with Yellow Feet
Cut maize, also called kibbled maize, provides pigments that will give really rich coloured shanks and feet. Maize is relatively low in protein and high in energy, which can lead to excess body fat, so maize should not be more than 10 per cent of the diet. Maize also helps to give rich coloured egg yolks.

Breeds with Yellow Feet and White Plumage
This applies to White Leghorns, White Plymouth Rocks, White Wyandottes and others. The problem here is that the cut maize will also give the plumage a creamy tinge instead of the pure white required, and the solution is to carefully time the feeding of cut maize. Start giving some maize towards the end of the chick crumb period, say from about six weeks onwards, but then stop the maize at about twelve weeks of age, before the birds start moulting from juvenile to adult plumage. Each adult year thereafter, give maize when the annual moult is completely

finished, and stop well before the next moult. This also applies to breeds with a plumage pattern over white ground colour, Black-Tailed White Japanese and Silver-Laced Wyandottes, for example.

Black- and Lavender-Plumaged Varieties

These pigments use a lot of the vitamin riboflavin, and if such birds suffer a shortage, the result will be poor feather quality, especially in the main tail feathers. Breeding stock with a riboflavin deficiency will produce chicks with 'clubbed' down. This will grow out, but it is not the best start in life for a potential champion. Use a suitable vitamin supplement; the types based on yeast powder are probably best for riboflavin.

High Grain Diets for Game Breeds

In an effort to achieve 'hard'-bodied birds, essential with exhibition Old English Game, breeders have traditionally used mostly whole-wheat-based diets. The fact that such a diet will prevent females from coming into lay until April, and not laying many then, is a bonus, not a problem, because even hens are expected to have a compact abdomen at a show. Many pullets will not have developed enough size for the November and December shows, so the exhibitor will concentrate on winning some prizes from January to March.

This is fair enough for nearly mature birds, but Old English Game breeders should not imagine that the nutritional studies done on poultry generally do not apply to them. A Game chick needs just as much protein and vitamins to grow properly as a broiler. They should be started on chick crumbs the same as

any other, even though the later changes in diet as the birds reach maturity may be different from other breeds. Old English and other Game breeds are required to be very physically fit, so must combine good muscle formation with activity. Many breeders use swing perches in aviary-type runs as 'gym equipment', because when a bird flies up to a swing perch, and it swings, they have to use all their muscles to maintain their balance.

Extra High Protein Diets for the Largest Breeds

In the Brahma, Cochin, Jersey Giant, Orpington and Wyandotte classes, to name but a few, the biggest birds will generally win – although plumage pattern and other breed points matter as well, of course. This may lead to problems of excess internal body fat, but the immediate aim is winning the cup, not producing a laying or table flock. In spring, after the main showing season is over, breeding stock is encouraged to run out and eat grass to lose some of that 'show fat' for improved fertility and egg production. Overly fat hens frequently lay long, narrow ('torpedo') eggs that are unlikely to produce healthy chicks.

Bantam Diets

Do not try to produce smaller bantams by poor feeding: some bantam breeds are frequently bigger than the official breed standard requires, but this cannot be corrected by poor feeding, which will only result in sick birds, or poorly fleshed birds that still have a larger than ideal frame. Size varies within a population, as do other characteristics, so the primary way to achieve a uniformly good flock is to breed a lot so you have plenty to choose from.

CHAPTER SIX

Incubation, Broody Hens and Incubators

Exhibition poultry breeders seldom achieve the success rates routinely achieved in the very large incubators used in commercial hatcheries. There are several reasons why this is inevitable, so poor results may not be 'bad luck' or because of incorrect methods. The first difference is the viability of the hatching eggs. Hybrid breeding stock is more vigorous than inbred exhibition stock, especially the short-legged and feather-legged breeds. Some novice hobbyists may hatch almost every egg in their small incubator if they are setting crossbreds' eggs, but when they later try an exhibition strain, they may wonder why fewer hatch.

Small incubators intended for hobbyist breeders are much more sophisticated than those used by previous generations of fanciers. Automatic egg turning and accurate electronic temperature control are now standard features on most models – but despite this, a fifty-egg incubator will not have all the features included in the latest 100,000-egg commercial machines; these 'extras' include automatic humidity control, unlikely ever to be available on models for fewer than 1,000 eggs. Further aspects to consider when choosing which incubator to buy are detailed below.

Broody hens are regularly used for hatching by hobbyists, and a reliable hen can be the best option with some difficult strains. An incubator is, after all, only a large imitation broody. Not all hens are reliable, of course – some give up a few days before the eggs are due to hatch –

but despite these problems, broodies remain a vital part of every hobbyist's breeding procedure. Many old poultry books devote considerable space to the 'problem' of broodiness, describing the methods to stop hens in laying flocks from sitting; however, most hobbyist poultry keepers today are more interested in ways of encouraging hens to go broody.

This is a potentially complex process so readers should buy a specialist book on the subject. A detailed description, with good illustrations, of the stages of embryo development, will help identify problems, as many usually occur at specific days.

Most fanciers in fact use both incubators and broodies. Hens seldom go broody before late April, but many breeders will hope to start their incubators in February in order to have some young birds ready to show in October. Those who specialize in breeds that seldom, if ever, go broody will obviously rely on incubators, but they often also keep a few Silkie-cross or other reliable sitters. Eggs with poor shell quality that would not normally be incubated, have more chance of hatching under a hen than they do in an incubator. It sometimes happens that the best show hens lay poor eggs.

THE IMPORTANCE OF EGG QUALITY

Only good quality eggs are likely to produce healthy chicks, or any chicks at all. Before the details of incubation are considered, whether this is under hens or in incubators, egg quality must be fully understood.

Apart from air passing in and out through the eggshell, a developing embryo has available only the nutrients present when the egg was laid, to grow into a fully formed chick. Thus it is much more important to ensure that breeding flocks are fed correctly, with no vitamin, mineral or amino acid deficiencies, than it is for eating-egg production.

Eggshell is quite a complicated structure. It is composed of several layers, with pores running through and two membranes inside. The overall shape of an egg is also vitally important, especially so for those species of bird that nest on mountain and cliff ledges: their eggs have very pointed ends so that they roll round the nest in a very tight circle. This is clearly not as critical with domestic chickens, which have been developed from ground-nesting jungle fowl. However, as the chick is getting into its hatching position, it is important that there is a distinctly different broad end and pointed end, or the chick may end up with its head at the pointed end, where it will be unable to hatch.

The crystalline structure of eggshell is similar to the wedge-shaped bricks of an arch. Because of this it is relatively strong when compressed from the outside – but a chick inside can easily push the 'bricks' out when it is hatching. Another factor that helps the fully formed chick to hatch is that the shell is less dense than it was at the start of incubation. This is because the developing embryo has been absorbing calcium from the shell to make its bones by way of the network of blood vessels that develop around the egg during incubation. The natural world seems not to waste anything.

As said above, there are two membranes covering the internal shell surface. At the broad end of the egg these separate to give an air chamber or 'airspace'; this is quite small in fresh eggs and gradually increases during incubation. Its size can be easily checked during this period by 'candling' – that is, shining a light through the egg. The internal humidity of the incubator can be adjusted if the airspace is too large or too small during incubation. Air can pass through the pores in eggshells to bring in oxygen and remove carbon dioxide while the embryo is growing. In the last few days before hatching, chicks move into the correct position and they peck through the inner membrane and start breathing air. Before this time oxygen was obtained via the system of blood vessels that quickly develop around the internal surface. There are more shell pores at the broad end of eggshell than elsewhere.

Bacteria and other harmful micro-organisms can invade eggs through these pores, but eggs do have some defensive capabilities in the membranes and albumen. Naturally incubated eggs are coated with skin oils from the broody hen, which act as 'natural antiseptic cream': this is a useful extra defence, especially if one of your best exhibition hens lays eggs with poor quality shells. When possible, it is a good idea to put such eggs under a broody for a few days (to get 'oiled'), and then transfer them to an incubator. The broody can then be re-started with more eggs from the prized but poor-producing show hen. It might be fairly argued that this system helps to perpetuate a natural weakness in an exhibition strain, but the poor shell quality might not be inherited, and anyway exhibitors will be most interested in producing another champion hen above all other priorities.

The future chick in a fresh egg is the 'germinal disc', a light-coloured spot about $\frac{5}{32}$in (4mm) in diameter, on the surface of the yolk. A network of blood vessels gradually spreads over the surface of the yolk. There are other membranes surrounding the developing embryo and these also have a system of blood vessels. When hatching eggs are regularly candled, usually on, or soon after the seventh day, blood vessels can be seen all around the inside surface of the egg. Specialized books on embryology and incubation should be consulted for more details of chick development, this subject being beyond the scope of this book. Such detailed reference books are essential for regular breeders, as they will be an aid to discovering the causes of incubation problems. Apart from the germinal disc, all the other egg contents are nutrients for the embryo, so it can be assumed that several of the potential problems during incubation are caused by nutritional deficiencies. It is therefore essential for breeding stock to have the best possible diet.

NEST BOXES AND EGG QUALITY

Eggs that are laid on the floor of the hen house or in the outside run, which sometimes happens, are a potential loss. They might get cracked by hens stepping on them, which will very likely set the birds off egg eating – and once hens start this vice it is very difficult to stop them. Eggs laid outside may be eaten by magpies, rats or other wildlife. If eggs laid anywhere other than a nest box are not broken, they will at the very least become much dirtier than eggs in a nest box, and although eggs do have defence mechanisms against bacteria and fungal spores, these defences can be overwhelmed if there are too many invading microbes.

Hens are naturally inclined to seek out a secluded nesting place. They still think they are wild jungle fowl, and will attempt to follow as much of their instinctive behaviour patterns as they can within the limits of a domestic poultry-keeping environment. Anyone who has kept a fairly large flock of layers in a traditional deep litter-type house (with or without an outside run) will have noticed that the most popular nests are those in the darkest corners. This should be remembered when choosing which poultry house to buy, or when designing your own DIY house. Also, if the nests are too high, requiring the hens to jump up on to a perch to enter them, and if the nests are opposite the house window, the hens might decide to find somewhere else to lay. If you have an existing poultry house with such unpopular high and light nest boxes, they can be made more attractive by fixing 'curtains' to partly cover each nest box. Hessian sacking was the traditional material for this, but you can use any suitable material available. If some hens persist in laying in their favourite

dark corners, there may be no alternative but to provide a box in the corner.

Nesting materials should be attractive and comfortable for the hens, of good cushioning quality, and cheap enough so that nest contents are likely to be regularly replaced. Hens will usually scratch about in a nest before they lay, and when actually laying they are usually standing in a squatting position, and not sitting; it is therefore very likely that the normal shavings, hay, straw or whatever will be scratched aside, exposing the hard nest floor so the egg will crack as it lands on it. The best solution is to cover each nest floor with a folded paper feed sack or newspaper, and put the shavings and so on, on top; thus there will still be a 'cushion' left after each hen has completed her scratching routine. You can really pamper your hens if you have an old carpet that can be cut into squares as nest lining.

This book has been written in Britain with European conditions in mind, but overseas readers might have alternative nesting and floor litter materials available, such as waste seed husks and other dried vegetable by-products from tropical agricultural crops.

With fairly large flocks of either deep-litter or free-range layers that are required to carry on laying and not go broody, it is normal practice to allow one nest box for every five hens. But breeding for exhibition as a hobby is a very different activity, and exhibition birds are usually kept in very small groups, one cock with six hens being a maximum for most breeders. Also, most hobbyists will encourage hens to go broody, not prevent them, therefore more nest boxes are provided, nearer to one nest per two hens.

COLLECTING EGGS FOR INCUBATION

Many readers will be surprised to see this heading, probably wondering how much anyone can say about picking up a few eggs. But there is a difference, because these eggs are intended to produce the next generation of your strain, they will probably be stored for several days before incubation is started, and there will be a lot of them.

Breeders should get in the habit of collecting eggs as often as possible, and writing the laying date and origin (in pencil, on the shell, gently!) of every egg. It is particularly important for specialist breeders to do this, as all the eggs will look much the same. Someone with three or four breeding pens of a single variety may have a fertility or hatchability problem with one of the pens, and will need to know which pen is the source of the problem. A fancier with several colour varieties of a breed – for instance Black, Exchequer and White Leghorns – and several pens of each, might mark their eggs B1, B2, E1, E2, W1, W2 and so on.

Eggs should be placed directly into cartons as they can easily be cracked in a coat pocket as the collector reaches down to secluded nest boxes in dark corners.

Frequent egg collection is vitally important during periods of very hot or very cold weather. Even fairly clean nests will be much dirtier than a clean egg carton, and less likely to harbour organisms that might invade and rot an egg in the heat of the summer. And in the winter, once an egg has been frozen it will never hatch, because the embryo is dead. As many breeders will be hoping to hatch some chicks in February, it is essential that every egg be collected each evening.

STORAGE OF EGGS FOR INCUBATION

Wild jungle fowl and pheasant eggs will be in the nest for up to twelve days before the hen starts to incubate them. Although eggs of this age do hatch, it is better to use eggs no more than seven days old. Eggs for incubation should be stored in a cool room between 10°C(50°F) and 18°C(65°F), preferably nearer the lower limit. Some slow embryonic development can start above 21°C(70°F), which is very harmful, so this must be avoided. If eggs have to be stored for over a week, their viable life can be extended by a few days if they are turned daily for the whole storage period. If

the eggs are in normal egg trays, the trays can be propped up at an angle on a block of wood, a brick or a thick book, first on one side, then the other.

The natural skin oil from a hen is probably a key reason why twelve-day-old eggs under a hen will hatch better than twelve-day-old eggs in a temperature-controlled storage area in a commercial hatchery.

Cleaning Eggs

Washing eggs, even if proper egg sanitant is used (available from suppliers of poultry keepers' equipment), is better than incubating very dirty eggs, but never as good as having eggs that were never dirty. If you have to wash eggs, it is best to do it under a running tap. Placing eggs in warm water is only likely to increase microorganism invasion.

FERTILITY OR INCUBATION PROBLEM?

Proper records of the breeding process should be kept in order to identify and correct any problems that arise. This is easily done on a postcard (or similar) for each broody hen or incubator batch of eggs, and can be kept handy. Full details of each batch of eggs started is noted, as is the number of infertile ('clear') and 'early dead germ' eggs removed when candled; and at the end, the number of live chicks is noted, along with a record of the 'dead in shells'. Specialist incubation books will give detailed information on probable causes of death, which will help identify any problems.

At candling it is important to correctly differentiate between infertile eggs, which will look the same as fresh eggs when candled, from early dead germs, which will show some development. Early dead germs often show as 'blood rings', where the network of embryo blood vessels has collapsed to show up as a single red line around the middle of the egg. Infertility is a breeding stock problem, possibly genetic, nutritional or management, but not incubation related. Dead embryos, whether early or late, might still be a genetic, nutritional or management problem, but might have successfully hatched if incubation conditions had been perfect.

SELECTION OF EGGS FOR INCUBATION

Most breeders will have more eggs available than they can incubate, so they can choose the ones most likely to succeed. Some factors to consider are eggshell colour and quantity, egg size and shape, and cleanliness:

Eggshell Colour: This is inherited, and is a vital characteristic of the dark brown and blue/green egg breeds. Barnevelder, Marans and Welsummer bantams often do not lay quite as dark-shelled eggs as their large fowl counterparts, so some improvement would be welcome.

Egg Size: The size of the egg has a direct effect on the size of the chick, and this effect can continue through to final adult size. Large-fowl breeders are therefore advised to select the largest available eggs for incubation, and those with a bantam breed that is often over-sized, should select rather small eggs. The size of adult birds is affected by many other factors (genetics, nutrition, housing and so on) apart from the size of the egg it was in when an embryo, but this is a real variable. Double-yolked eggs never hatch.

Egg Shape: This has a major effect on chick viability. The ideal is a 'classic' egg shape. Long, narrow, 'torpedo' eggs do not have enough space for an embryo to develop properly, and almost spherical eggs, with no obvious pointed end, often do not hatch because the chick is not in the correct position, with its head at the air-cell end.

Eggshell Quality: A good eggshell is needed to protect the embryo during incubation. Porous shells – these feel rough textured, and are often worse at the pointed end – 'tramlines' on shells, and similar shell faults, are not worth incubating.

Cleanliness: A washed egg may be better than a dirty egg, but it is better if the eggs were clean to begin with. It is best to wash eggs under a running tap, as immersion in dirty water may do more harm than good.

SOME CAUSES OF INFERTILITY

The causes of infertility are many and various, and include inbreeding (genetic), disease, an inhospitable environment, nutrition, and social and psychological problems amongst certain groups.

Genetic

If a strain has been inbred for several generations a number of harmful recessive genes may have come into operation, which may well affect the production of viable sperm or ova. The only option is an out-cross to another strain, one that must be carefully chosen to retain as much breed standard quality (plumage pattern and so on) as possible.

There is one specific gene that should be mentioned, and that is the rose-comb gene. For reasons I have never found an explanation for, the males of any rose-combed breed which are pure (homozygous) for rose comb have poor sperm quality and quantity. This problem does not affect males that are rose-combed but are carrying the (recessive) single-comb gene, nor does the gene affect female fertility in any way. Because the males that are carrying the single-comb gene are fully fertile, and cannot be identified visibly from 'pure' rose-combed males, the single-comb gene is perpetuated in these breeds. Inevitably some of the breeding hens will also be carrying the (unseen) single-comb gene, and a few single-combed chicks will be bred. Such single-combed chicks are of no use as breeding stock and are usually killed or, if reared, sold as pets/layers.

Disease, Past and Present

An obviously sick bird of either sex is not going to breed. Less obvious conditions can be just as harmful, including internal or external parasites. Some conditions can leave permanent infertility, even though the bird appears to have fully recovered and looks perfectly healthy. One example of this is infectious bronchitis, which can permanently cause eggs to have 'watery whites'. Affected eggs may get as far as 'early dead germs', but no further.

One or more specialist books on poultry diseases will be needed to even begin to appreciate the number of diseases which might affect fertility.

Environment
The rain and inevitable mud of a British winter, or anywhere else with a similar climate, is certainly not the right environment to get chickens 'in the mood'. Despite the fact that domestic chickens have been bred in Britain and the rest of central and northern Europe for over two thousand years, and in that time have been developed into a host of local breeds, in some respects they are still Indian jungle fowl. Leading exhibition breeders may be providing much more luxurious and specialized accommodation than other poultry keepers, but very few of them are able to provide a completely controlled environment. Probably the nearest to this level of control is in the winter housing used by some enthusiasts in Canada, the northern states of the USA, The Netherlands, Germany and the Scandinavian countries.

Basic hygiene must be considered. Dirty conditions increase the chance of diseases, and all the problems that come with them. This should never be a problem with exhibitors, who, if they are to be successful at shows, should have much higher management standards. One problem that may occur is reduced laying and fertility among birds housed inside in very hot weather. The effects of the heat can be worsened by ammonia gas given off from the floor litter.

Breeders of ducks, geese and turkeys should be well aware of their special requirements. Geese and the heavier breeds of duck normally mate in water, so a pond must be provided. With turkeys, one of the main 'environment' requirements is the provision of canvas saddles for turkey hens, to prevent the stags from injuring their backs when mating.

Nutrition
Correct nutrition is described more fully elsewhere in this book, so here suffice it to say that deficiencies of any nutrients, especially some vitamins, can adversely affect sperm quality and, most critically, the nutrients in the eggs which will be needed by the developing embryo.

Preferential Mating
Cockerels and drakes are renowned for their virility and willingness to mate with any available female; however, the reality is not always the same as the myth. Light breed cockerels (such as Anconas) and drakes (such as Khaki Campbells) will usually live up to the legend, but some heavy breeds, such as Cochin chickens and Aylesbury ducks, will be much less active and will effectively bond with just one or two females, ignoring any others in the pen. This tendency to form close and exclusive relationships is even more pronounced with geese and certain breeds of chickens; for instance, Sebrights and the Asian hard-feather breeds (Asil, Malay, Shamo and so on).

Poultry behaviour experiments have found that chickens will tend to prefer either (a) their own breed, or (b) birds of natural jungle fowl black-red/partridge plumage colour.

Abnormal Imprinting
Broody chickens and bantams are often used to hatch and rear ducks, guinea fowl and pheasants, and this practice can result in some very confused adult birds. To minimize these problems, the 'adopted' chicks should be reared within sight of their own species, and mixed with them as soon as they are old enough to leave the hen.

Bullying and Psychological Castration
Subordinate males in a flock will be intimidated by larger and stronger males, and this can have a long-lasting effect; thus a smaller cockerel in a rearing group may lack confidence later in life even when put with some hens of his own in a separate pen. This is another good

reason to cull sub-standard, growing cockerels as soon as their faults are apparent.

Large breeding flocks with several males and a host of hens all running together, are not as idyllic for the chickens as some people might imagine. Subordinate cocks are literally knocked off hens by the dominant males when they are trying to mate, and may eventually give up.

Bored and Hen-Pecked Husbands

The social behaviour of chickens bears a greater similarity to that of humans than some people might imagine. Settled groups, such as geese and the preferential mating breeds mentioned above, should be left alone, but some males, especially older cocks, may lose interest in their hens and mope about. Some hens will take advantage and start to peck at the cock's feathers or any scabs they see on his ear lobes.

Most exhibitors will have several males of each variety, with the 'spares' being housed individually. If the cocks are swapped around every few days they will stay 'keen', and not allow themselves to be 'hen-pecked'.

INCUBATORS

Choosing an Incubator

At the risk of sounding like an incubator salesman, it is true to say a really good model, with automatic turning and the latest electronic controls, is worth the money. To be more specific, features that should be to considered are capacity, the outer casing, how easy it is to wash out, and the availability of parts, and the after-sales service.

Capacity: Many more eggs will need to be incubated than many beginners imagine. This is especially true of those breeds that are well known for poor fertility, such as Indian Game (=US Cornish). Even with more fertile breeds, a lot of chicks will be needed to obtain enough good quality adult pullets.

Very small models are sometimes less able to maintain a constant temperature than larger models, despite the electronic controls. This

is because the temperature changes in the room where the incubator is running are too great for the heater to cope with. The author recommends a minimum forty hen-egg size machine. If infertility problems are anticipated, it may be better to have two medium-sized incubators rather than one very large one: this is because incubators hold to the correct temperature better if they are at least half full. If, say, two fifty-egg incubators are started together, when all the clear and early dead germs are removed at candling, the viable eggs might all fit in one machine.

Insulated Outer Casing: Models with a thick insulated outer structure are likely to give better results than those of thin plastic. Most designs include an inspection window, which is good, but does the whole thing need to be transparent?

Washability: It is vital for the inside of an incubator to be thoroughly washed between hatches. Some designs are easier to take apart, wash and reassemble than others.

After-Sales Service and Spare Parts: The availability of spares and advice should be checked, especially if the incubator is made in another country. Unless there is good reason for not doing so, the best choice will be the incubator with a local service centre.

The Management of Incubators

- Follow the instructions provided!
- Run the machine in the place with the least fluctuation in temperature available. This should also be somewhere where the incubator will not accidentally be knocked or kicked by anyone in the family (including the dog) or slept on by the cat, as this will affect the running temperature. Be aware that the environment may affect the amount of water needed in the incubator to maintain the correct humidity for incubation and hatching. A cool but even temperature in a cellar, or a warmer, drier, centrally heated

room might both be suitable places to run an incubator, though the machine will need to be set differently.

- When the chicks start to hatch, do not start removing them to their heat lamp, but leave them where they are for twenty-four hours. A high humidity level is essential to keep the egg membranes soft at this critical stage, and opening the incubator will let all the warm, damp air out. The chicks have enough nutrients from the remaining yolk to keep them happy for this period. Once the main group of chicks has been removed, the incubator can be left running for late chicks. Empty eggshells should be removed, as they can cap over eggs that have yet to hatch.

Egg Candling

All eggs being incubated, whether under hens or in incubators, must be tested or 'candled' at about the seventh to tenth day, or rather night, of incubation. Proper 'egg-candling lamps' can be bought for the purpose, or home-made ones used (a light bulb in a wooden box, with an egg-shaped hole on one side), or the eggs can simply be held over a torch (= US flashlight), with the hands so positioned to only let light out through the egg.

Infertile ('clear') eggs will appear the same as fresh eggs, just showing the undeveloped yolk as an orangey glow. The embryo will show as a dark spot, and its network of blood vessels will also be visible in all but some dark brown or olive-shelled eggs. Embryos that have started to develop, but have died after the first few days, will be smaller than the growing ones and the blood vessels often collapse into a 'blood ring'. Experience will make this job easier. It is important to remove clear and early dead eggs as they rob heat from the growing embryos and may go rotten, sometimes even exploding. They are a dangerous source of harmful micro-organisms.

BROODY HENS

Choosing Broody Hens

Those who keep breeds that go broody will naturally just use these hens, but breeders of non-

sitting varieties often buy in hens especially for the purpose. Have broody hens of approximately the same size as the breed being bred, since large broody hens may crush tiny bantams, and tiny bantam hens will not be able to cope with very large eggs. At the small end of the spectrum, Old English Game bantam hens that are not good enough to show can usually be bought cheaply, and are reliable sitter and gentle mothers for the tiniest eggs.

Silkies are renowned for their abilities as broodies, but their silky plumage can be a hazard to young chicks, as they can get tangled in it. This is why part-Silkie crossbreds are more popular. If Gold Silkie males are crossed with silver-plumage females, Light Sussex (large or bantam) or Silver Duckwing OEG, the resulting chicks can be sexed at a day old. Both large and bantam OEG, and crosses, are also popular as broodies because they are good mothers and have a lot of heat-generating breast muscle, ideal for incubating eggs through still, chilly spring nights.

The Management of Broody Hens

A hen going broody will be noticed still in a nest box overnight instead of on her normal perch. Do not be in too much of a hurry to move her to a separate brooding cage as she may go off the idea. In normal circumstances, with the eggs being regularly collected, she may not be sitting on very much at this stage. If it is intended to encourage hens to go broody, it is a good idea to leave at least two dummy eggs in each nest at all times. For bantams and some of the smaller breeds of large fowl, old golf balls may be a cheaper alternative. When a hen appears to be going broody, the first step is to give her a few more dummy eggs/golf balls/real eggs (but not the ones you want to incubate) to keep her interested. This should be done in the evening when she is half asleep, and therefore less likely to go running and squawking off into the distance.

Once a hen is fully broody – and this involves significant hormonal changes – there is a much greater chance of moving her without upsetting her. Broodies cannot be left to sit in a breeding house with other hens still present: they will disturb her, and her eggs, by laying fresh eggs in the nest box. The changes from normal to broody behaviour are caused by prolactin from the pituitary gland. General metabolic rate drops, as does food consumption and (obviously) general activity. It is normal for hens to gradually gain bodyweight with age, and this may partly be an evolutionary refinement to provide a fat reserve when they are broody. If the nest is approached, the hen reacts in a defensive and aggressive way, very different from her normal response, which would be to run away. A lot of breast feathers fall out (and may become part of the nest lining) to reveal the hot skin of the breast muscle. Many novice poultry breeders make the mistake of associating profuse plumage with broody quality; however, feathers only hold heat in, they do not generate it. Old English Game hens have long been valued as one of the best broody breeds, and this is because, although they have fairly sparse plumage, they have very well developed breast muscles, which are the real source of body heat. Once the chicks have hatched, hens normally have a complete moult, partly to replace the lost breast feathers, and partly to lose any external parasites that may have been picked up during sitting.

While hens are sitting, they instinctively control the temperature of their eggs by moving the eggs around, so all of them are evenly heated. They will leave the nest to feed, more so towards the end of incubation, by which time their reserves of body fat will be depleted. Embryos can withstand some cooling time, and more so towards the end when they will be generating some body heat of their own.

As said above, broody hens need to be kept in a cage on their own. Standard-size show cages for large fowl (24in/60cm cubes) are ideal for broody bantams and smallish large fowl, but larger ones will be needed for big hens. This will accommodate the nest box, and leave enough space for the hen to eat, drink and defecate. The nest should be sprayed with a mite killer before the hen starts sitting;

however, it may still become infested by hatching time. The most hygienic nests are therefore suitable cardboard boxes that can be used only once, or washable plastic boxes made by cutting a hole in the side of a 5gal (23ltr) drum. Your home/work/life might provide some other suitable disposable containers. The ideal size will naturally depend on the size of your broodies: it should be large enough for the hen to sit comfortably on her nest, but not so large for eggs to roll out. If your box is rather large, this problem can be solved by putting a brick in at one end to keep the eggs in.

Some breeders use open-topped boxes, but I don't like them because they are not dark enough for hens to feel secure, and they can break eggs when they jump in. Also, nests that are entered from the side can be covered to keep the hen shut in for the first night or two while she is settling in to her new nest. If disposable cardboard or washable plastic nests are used in the breeding pens, the hen can be moved to the brooding cage in the nest in which she has already settled.

Once moved into a suitable cage for brooding duties, the hen should be left with her dummy eggs for two more days to be sure the move has not put her off the idea. If all is well, in the second evening replace the dummy eggs with the eggs to be hatched. On the seventh or eighth evening, use a torch to test the eggs. The developing embryos and their network of blood vessels will be clearly visible, except possibly for eggs with very dark brown and olive (from some Araucanas) shell. Remove all obvious clears, because they can go rotten; and even if they don't, they rob heat from the live embryos.

When hatching day arrives it may be safest to shut the hen in her nest box to prevent chicks from falling out of the nest and becoming chilled. It would be a shame to lose them now. However, do not constantly disturb the hen to have a look at how things are going: once every six hours is the maximum. When you do check, gently remove the empty eggshells, as they can cap over eggs that have not yet hatched.

CHAPTER SEVEN

Chick Rearing

This book is entitled *Exhibition Poultry Keeping*, and so the author assumes that most readers will already have some *How to keep ...* books, and will have bred some chicks before. Therefore this chapter is fairly brief, and concentrates on the special needs of show breeds.

REARING WITH BROODY HENS

If more than one hen has been sitting at the same time, and some have only hatched a few chicks, any movement of chicks from one hen to another should be done as soon as possible, preferably before the families are moved to rearing pens. Most owners are familiar enough with their birds to know which hens are likely to accept chicks.

Most established fanciers will have several different-sized cages and pens for various ages and sizes of bird. Standard large fowl show cages (2ft/60cm cube) with a nest box in a corner are ideal for sitting hens, but larger premises with chick-proof sides will be needed once they hatch, and further moves to even larger pens as the youngsters develop.

The traditional housing for a hen and chicks is a 'broody coop and run' that is moved every day around a lawn, and no doubt many thousands of chicks have been successfully reared in these coops; however, they are not the author's preferred system. Very young chicks have several very basic requirements: warmth, food, safety, and freedom from disease organisms. Fresh air and green grass are beneficial when they are older, but not for the first few

weeks. Well constructed, raised cages inside a shed are warmer, drier, safer from rats and other predators and much less likely to bring the chicks into contact with disease than outside runs. Wild birds spread avian diseases in their droppings, and disease organisms, especially *Coccidia* spores, can survive in the ground for a very long time; damp conditions outside are much more favourable for the *Coccidia* than for the chicks.

These cages should be at least 2ft (60cm) high, 2ft (60cm) deep, and 4ft (120cm) long. It is assumed that a block of cages will be built against the wall of a shed. Ideally all internal surfaces should be smooth and washable, of painted or melamine-coated board. The fronts should be of small mesh, to safely keep the chicks in, and any mice, rats and sparrows out. If the cage is larger, the family can stay in it longer before they need to be moved on. The exact size of a pen will naturally depend on the size of the shed in which it is constructed. Each cage should be thoroughly cleaned before the hen and chicks are moved in, and the cage floor should be covered with a thin layer of wood shavings (a deep litter will seem like a mountain range to day-olds!). More shavings can be added gradually over the next few weeks.

Water should be provided in smallish-sized plastic founts: the smallest size is ideal for a bantam hen, but the next size up is better for large hens that may kick over small drinkers. The largest drinkers have deeper troughs, possibly too deep for some chicks to reach. Chick crumbs should be given on the cage floor for the first few days. Hens can get very excitable at this stage, and troughs that are shallow enough for very young chicks to reach or jump into, may be light enough in weight for the hen to throw over. Chicks have been killed by being trapped underneath overturned food troughs, and this danger must be remembered when choosing a suitable trough for the next few weeks. Heavy and shallow pottery bowls might be found at your local car boot sale, and these are better than the purpose-made but light-weight troughs sold at agricultural suppliers.

Hens with chicks can be moved to outside arks with runs, if preferred, once the chicks are about a month old. Runs with solid (rain-proof) tops are best. Because of the risk of infection, it is suggested that arks for chicks should be on fresh mown grass not used by adult birds. Many fanciers have a poultry part of their garden, and a 'normal' garden: if the rest of the family will agree, hen and chick arks might be allowed to 'invade' the family lawn every May and June so as to get the young birds off to a good, safe start.

Many paintings depict hens with their chicks free ranging around farmyards. In the real world this is seldom a good idea, however, as chicks can easily be lost or fall victim to pred-ators, including birds of prey.

Growing chicks are old enough to come away from their mother at about eight weeks of age. The exact age will vary according to the breed, in that some develop more quickly than oth-ers, and it will also depend on the personality of the hen: some start to get aggressive with the chicks when they think it is time for them to go. If the next stage housing is to be a house with an outside run, it can be a good idea to move the family, still with the hen, shortly before the hen is due to be removed; she can then help introduce the growers to their new surroundings.

REARING CHICKS UNDER LAMPS

The same considerations of warmth, safety and protection from disease are even more impor-tant for lamp-reared chicks than those under hens. In order to keep the rearing area warm enough, the 'pen' they are in may need to be more enclosed, and the sides are normally solid all round. For small groups, a very large card-board box with a floor area of about 30 × 36in (75 × 90cm) is ideal (such boxes might original-ly have been used for larger household appli-ances, lawnmowers). A lamp must be suspend-ed over the box: larger groups will need a proper infra-red light bulb, but small groups can be reared using a normal 100-watt light bulb. Screw-fitting bulbs must be used, since even normal light bulbs should be fitted in heat-resistant ceramic holders (these can be bought at any agricultural supplier). The rearing box should be set up well before the chicks are due to hatch, and the lamps switched on the day before they are due so that it will have warmed up ready. Buy at least one spare lamp, so the one in use can be replaced immediately if necessary.

Experience and chick behaviour will show if the lamp is at the correct height: if it is too high, the chicks will be cold and huddled together directly under it; if too low, and hence too hot, they will be as far away from the lamp as they can get. Obviously a 100-watt light bulb will need to be lower than an infra-red lamp. As the chicks grow they will need less heat, and so the lamp can be raised and/or changed to a lower wattage. At about four weeks of age they can be moved to a wire-fronted cage in a shed, as described for hen rearing, though still with a light bulb in the cage.

When setting up for day-olds, the major dif-ference between lamp and hen rearing is the floor covering. A thin covering of wood shavings is advised when a hen is used, but under a lamp the floor should be covered with coarse materi-al – an old hessian sack or piece of carpet is ideal. This is because the lamp-reared chicks do not have a hen to show them that chick crumbs are for eating, and wood shavings are not, and lamp-reared chicks have killed themselves by eating shavings. Shavings can be added after the first week, by which time the chicks will have learned about food.

Small-size plastic feeders are fine for heat lamp-reared chicks, as they will not kick them over like broody hens. For the first few days crumbs should be sprinkled over the pen floor as well, so they can find it. Some chicks, especially turkey poults, can be very stupid, and if a batch of chicks does not seem to have started eating and drinking after a few hours, it may be necessary to show them by making pecking movements with your finger and gently dipping their beaks in the water.

LATER STAGE REARING

The chapter on nutrition (Chapter 5) should be studied, with particular attention to the slower growth rate of pure breeds as compared to commercial hybrids, and the 'Special Diets for Exhibition Breeds' section. Many fanciers will be rearing mixed groups of different breeds, possibly both large fowl and bantams. This is usually fine for the first few weeks, but gradually problems will arise: thus small bantams will be pushed about by large growers, boisterous breeds by quiet, docile breeds, and cockerels may start to fight. It is part of good stockmanship to notice bird behaviour, dividing groups when necessary. If there are several groups of nearly equal ages, one under a lamp and perhaps two groups with hens, they can all be rearranged at the ten-week-old stage according to size and/or sex. The details of how to

divide them will be different for each flock. For example, it might be best just to separate the large fowl cockerels, leaving the bantam cockerels with all the pullets, for two months at least. Eventually all the cockerels may have to be housed individually, but for ease of daily care, only separate birds when necessary.

The breastbone of young chickens is composed of soft cartilage, not solid bone, a characteristic that can be seen on broiler chickens as bought from your local supermarket. Bent breastbones are caused by allowing young birds to perch too early – though eventually their natural urge to perch will have to be allowed. It is best to provide very wide perches at first, so their bodyweight can be spread over the whole length of the keel.

The period from eight weeks until about fourteen weeks of age is very important for breeds that will need to be kept clean, dry, and out of bright sunlight when adult in order to keep them looking good for showing. Growing birds can have a few weeks out on grass before they have to be shut in to keep the foot feathering and white plumage as they should be. As soon as the growers start to grow their first adult plumage, show preparation effectively begins. White-plumaged varieties with yellow feet (Leghorns, Rocks and Wyandottes) will yellow up nicely in this period, and the colour

A group of growers, mostly Light, Silver and White Sussex, housed inside. The heat lamp, seen at the top of the photo, will have been switched off by the time they are this old.

will stay for a long time after they have been housed inside.

LEG RINGS

Closed, numbered leg rings are compulsory for exhibition poultry in most European countries. At the time of writing (2004) they were voluntary in the UK, but it was expected that the British government would eventually require the PCGB to make leg rings compulsory here as well, as part of plans to improve traceability in case of disease outbreaks. These rings are issued by each national governing body in various sizes to suit males and females of each breed. When they are sent out, a record is kept of the ring numbers issued to each fancier, although there is no need to return details of which bird each ring was put on. Ring sizes for each breed are provided in a PCGB list, and are included in Dutch and German standards books.

The rings are fitted by sliding the ring over the three front toes, then the ball of the foot, and finally up the shank and hind toe. The exact age at when this is done is variable, about eight to twelve weeks; fortunately, the hind toe is very flexible at this age. Pullets have smaller rings than cockerels because as adults they have thinner shanks; however, at this age there will not be much difference. Thus rings have to be fitted on pullets a few weeks earlier than cockerels. Special care may be necessary when fitting them on feather-legged breeds. First-time leg-ring users will probably start trying to fit rings before the legs are thick enough, just to be on the safe side; however, if the rings are going to fall off, leave it for another week.

Leg rings on cockerels need to be constantly checked over the next month or two, to be sure that when the spurs start to grow, the rings can be slid up over the spur. This may not seem important at the time, but when the bird is a three-year-old mature cock, the lower shank (between the spur and the foot) will have thickened considerably, and a ring stuck there will be uncomfortably tight.

Removable coloured leg rings, put on the other shank, are used by many breeders in addition to the closed numbered rings. They will buy a collection of these in different colours and sizes, perhaps starting with very small-sized rings long before the fixed rings can be fitted. This is done as part of a continual assessment of the growing birds, which all breeders should do, and the growers can be given a certain colour for 'excellent', 'good', 'fair', 'poor', so they can be identified at a glance. This can be very useful if an unexpected customer calls, as he/she can then be told that 'All the birds with green rings are for sale', or whatever; it is difficult to start closely checking leg-ring numbers in front of a potential buyer.

DAYLENGTH AND GROWTH

Commercial rearers with windowless houses can control the daylength given to flocks of growing pullets. This is a facility that hobbyists do not have, and therefore they have to time hatching to fit in with natural daylength changes. A gradually decreasing daylength lengthens the growing period; increasing daylength induces sexual maturity and laying.

Exhibition large fowl need to be as large as possible, whereas bantams usually need to be fairly small, although some breeds can be undersized. Maximum size is usually achieved in chicks hatched from February to early May, because they have the optimum time of decreasing daylength at relevant ages, before they begin laying the following February. Chicks hatched in the winter – December and January in the northern hemisphere – sometimes start laying in July or August, and stop growing as a result. Similarly, late chicks, hatched in June or July, have to cope with cold weather when they are still young from October onwards, and then also start laying in February, the same time as their elder sisters, who have had an extra three months of summer in which to develop. Bantam breeds that are oversized can have their growth checked in this way, possibly getting two generations in one year, by hatching in November and hopefully again in July.

CHAPTER EIGHT

Show Bird Training and Preparation

This chapter should make it very clear that poultry showing is more a type of 'Folk Art' than a branch of agriculture. The idealized colour plates of breeds seen in late nineteenth- and early twentieth-century books, and given with poultry magazines, were intended to provide a template for breeders and judges. Art experts might criticize these pictures for being 'wooden' or 'chocolate box', and they would probably be right, but they were produced for a specific purpose. Then the real birds would be expected to look exactly like their picture: perfectly clean, perfectly coloured and patterned, and even standing in the same pose. A chicken just caught up from a farmyard and put straight in a show would not do. Many thousands of men who otherwise led very hard-working lives

spent all their very limited free time devoted to this and other small livestock and gardening-based hobbies. Any eggs produced were a bonus, and not the main motivation.

Two old exhibition breeds that received a lot of time-consuming special treatment were Malays and Spanish. These breeds were developed to show perfection rather closer to home than their names suggest, Malays in Cornwall and Spanish in and around Bristol. The Malay breed standard requires 'Head: Very broad with well projecting or overhanging (beetle) eyebrows, giving a cruel and morose expression' and 'Eyes deep set'. This expression would be improved by repeated sessions of massage. The birds would sit on their owners' lap, and the skin between the eyes and the comb would be

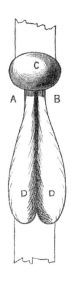

'Cock-saddle' used by Mr Roué of Bristol, circa 1850–70, when preparing the white faces of his Spanish for the shows. Specialist breeders of this variety led the way in preparation techniques when poultry shows started in the mid-nineteenth century.

gently massaged and stretched so that it would droop over the eyebrow bone to increase the glowering expression. Spanish fowls have unique white faces, and the area of loose, floppy white skin should be as large as possible: this was also increased by many sessions of gentle massage. No doubt the birds receiving this treatment soon became very tame, but for the first few sessions some breeders made 'saddles' to sit the birds in. No television then, of course!

GETTING BIRDS TAME

Judges vary in their tolerance – or intolerance – of 'wild' chickens at shows: some will quickly declare such birds 'unjudgeable', others will be more patient. If it is a small class, and none of the birds is well behaved, judges have to be patient; but wild birds may miss out on a prize in large classes if they do not stand quietly.

Chicks should be picked up and handled as often as possible once they have started growing well, say from two or three weeks of age. This is easier with incubator-hatched and heat lamp-reared chicks than with chicks under broody

'*A Well Appointed Training Cage*' from W.W. Broomhead's The Management of Chickens, *1913.*

hens. Try to find time to pick up at least one chick from each group every day; even those that are not picked up will be getting used to your hand in their pen. At the same time, you will be continually assessing the quality of the growers. There are usually more good, or at least acceptable, quality cockerels than can be kept through to adulthood, sold, or even given to 'good homes', so obvious duds might as well be killed as soon as they are identified as such.

Long-legged tall breeds such as Malays and Modern Game should obviously stand upright to show their height; at the other extreme, Japanese and Pekin bantams should be sedate, with their heads rather low to the ground. Training sessions will be useful, and should have a clear objective. However, the same also applies to most other breeds as well, and even very 'normal' breeds have a required pose. Rhode Island Reds and Sussex should both stand with a horizontal back, and they will only do this if they are happy and relaxed at a show. Excited birds will probably be in every position except that required, and frightened birds also tend to tighten their plumage close to their bodies – not good on profusely feathered Orpingtons or Wyandottes. Some breeds are naturally more tame than others; thus some of the light breeds, such as Anconas and Hamburghs, are the most difficult to tame, and these nervous types will need more time and attention – and even then will seldom be as confident with people as Ko-Shamo, which have been kept as pets in Japan for centuries.

Victorian and Edwardian exhibitors were even advised to have a selection of the ornate ladies' hats fashionable at the time, and to wear them when feeding their flock so that the birds wouldn't panic when genteel ladies visited the shows. I wonder if many of them, gruff farmers or stiff-upper-lipped gentlemen, could bring themselves to follow this suggestion?

HOUSING DESIGN AND SHOW PROCEDURES

Birds in raised show-training pens and permanent houses are likely to receive more human contact than those down at ground level, with adult poultry keepers anyway. Therefore if you have a choice, put those birds that are most in need of personal attention in the pens where they are most likely to get it.

Tall breeds should be given their food and water in high-level containers. Many years ago I bred Malays, and used hook-on 'D' pots for single show birds. One day I had to rush the feeding to get off to work, and threw the food into some of these pens. One Malay cock was very upset about this, and I noticed him looking in his (empty) food pot, and then looking down at his food on the pen floor, clearly thinking 'You don't expect me to eat it down there, do you?' – so I had to go back to feed him 'properly'. The front of pens for these tall breeds can also be made solid, up to a height where the birds will need to stand 'on tiptoes' to view their world.

With this in mind, it is obvious that, on the other hand, Japanese and Pekin fanciers will use pots on the pen floors – cat bowls from your local pet shop will be ideal. Special food and water containers are also used for Polands and other crested breeds, especially for their water. There are several ways of allowing these birds to daintily dip their beaks into the water without getting their crests wet, which they would do if normal open bowls were used.

Judging Sticks

A few judges have impressive antique sticks, but most use extendable lecturers' pointers. They are used to persuade birds to stand as required in their show cage, and it will be helpful if your birds have encountered one before they are shown. If they associate being moved around with a judging stick with extra tasty treats, so much the better.

Special Show Feeding

Normal nutrition is covered elsewhere, but some exhibitors have more or less 'secret' extra conditioning mixes. Most of these are intended for pigeons or cage-birds; some even use insectivorous bird mixes. Because of the cost of these foods, they are mostly used for very small bantam breeds. Heavyweights such as large Orpingtons would need a lot of bird seed to make any difference to their condition, and the seeds are so small they may not even eat it. Canned cat- or dog-meat is a popular treat, and gives some extra protein. The cockfighting fraternity has been keen on 'secret recipes' for centuries; I am not sure how good these mixes

'*Using the Judging Stick' from W.W. Broomhead's* The Management of Chickens, *1913.*

'*Training an Exhibition Bird to show his good points' from W. Powell-Owen's* How to Win Prizes with Poultry, *circa 1912.*

were, or even if half the ingredients are available, but they make for interesting reading in the game-fowl literature.

NORMAL PREPARATION, OR FAKING?

Washing birds for a show is normal preparation, and resorting to dyes or major cosmetic surgery would be classed as faking. Novice exhibitors will need to know the dividing line between the two. A lot more preparation is allowed than many people expect – and indeed, a lot more preparation is actually required by judges than many novices imagine is possible.

Dastardly deeds in the poultry-showing world were 'exposed' in George Riley Scott's book *The Art of Faking Exhibition Poultry*, published in 1934. A facsimile edition was published in more recent years, which may be easier to find for anyone wishing to learn more about the 'good old days' of poultry showing. Incidentally, this book was prominently featured in *Bizarre Books* by Russell Ash and Brian Lake in 1985, which accurately said: 'The author treads an indistinct line between condemning this widespread and despicable practice, and telling the reader exactly how to do it.'

Mr Scott mentioned – and he was not the first to do so – one of the most remarkable types of faking, and one which almost became 'normal' among a group of specialists. Silver Spangled Hamburghs have a very demanding standard, which is difficult to breed correctly: white ear lobes, comb shape, general body shape and carriage, and the spangled plumage pattern. The tail feathers, which should be clear white except for a large circular black spot at the end, are probably the rarest part of all to have anywhere near perfect. When breeders in the old days were lucky enough to have such feathers they were determined to make the most of them so, using skills learned from falconers, they transplanted good feathers into the cut-off quills of the feathers of inferior markings that were really growing on otherwise good show birds. Perfect male Pencilled Hamburgh tail feathers are just as difficult to obtain, and so may have been transplanted as well. These feathers are solid black except for a narrow edging of gold or silver, and this edging is the difficulty.

Silver Spangled Hamburgh bantam male. A complex variety that has been the subject of keen competition since well before the first 'organized' poultry show in 1845. Some over-enthusiastic showmen have been tempted to go beyond accepted preparation into cheating; the 'transplanting' of well-spangled tail feathers was the most common fraud.

Adding feathers obviously counts as cheating, but removing them is normal show preparation. A surprisingly large number of small body feathers can be removed without being noticed – although it would obviously be cruel to do many at once. You cannot pluck birds alive!

If any large wing or tail feathers are the wrong colour, a solid black feather on a barred variety for example, it either has to be removed at least two months before a show or not removed at all. There is no guarantee that the new feather that will grow back will be any different from the first one. In the case of a barred or cuckoo (coarse fuzzy barring) bird with white, or partly white feathers, these are less likely to produce better new feathers unless the reason for the original lack of pigment was stress or nutritional shortage during their previous moult. Growing a complete set of feathers is very demanding, and it is very common for black, blue, black-red and other dark-coloured varieties to have some white patches. If only one or two feathers have to be replaced a month or two later, they might get it right next time round. This is not the case with black mottles, however: these, and other plumage varieties that are particularly well known for selective feather removal, are detailed below.

Black Mottles: This colour variety, seen at its best on Anconas, has progressively more, and larger, white spots with each successive moult. Most Anconas have the correct amount of white in their first adult year, and are too white (or 'gay') in future years. Some birds have too few white spots first time round. Expert exhibitors know that if a few carefully selected body feathers are removed as soon as they have grown their first adult feathers, the replacement feathers should come out with proper white spots.

White-Crested Black, White-Crested Blue and White-Crested Cuckoo Polands: The crests are obviously white, but the front part of the crest is the same colour as the rest of the body. Some Polands have some coloured crest feathers, which if pulled out have a better than evens chance of coming white next time.

Lakenvelders and Vorwerks: These two breeds have solid black necks and tails, to contrast with white bodies on Lakenvelders and buff bodies on Vorwerks. The solid black necks are seldom bred, and if they are they are usually accompanied by black body feathers. The removal of a few feathers can dramatically improve many specimens, but some are too bad for this to work.

Campines: Silver Campines should have clear, silvery-white necks contrasting with the rest of their plumage which is heavily barred with black. Gold Campines have gold necks and black over gold barring. Birds with naturally clear gold or silver necks usually have poor barring on their breast, and there is nothing you can do to improve this. Therefore Campine breeders have tended to breed from well-barred birds, and then pluck out the few black neck feathers that usually followed. Campine males are 'hen-feathered'. Compare pictures of them with pictures of the closely related, but normal cock-feathered Brakel. Some Campines do produce a few silver or gold saddle feathers, which

White-Crested Black Poland bantam male. Note the (correct) black feathers at the front of the crest. The removal of incorrect coloured feathers to obtain the desired result is considered normal show preparation practice.

Lakenvelder female. Ideally all the neck feathers should be solid black. If there had been just a few white feathers, they could (legitimately) be removed for showing; however, this one has just too many white feathers for this to be possible. Lakenvelders remain very rare because not many breeders are willing to breed the large numbers necessary to get a few good ones.

Silver Brakel bantam male. This breed has normal cock feathering. Compare it with the hen-feathered Campine shown in 'Choosing Your Breeds', Chapter 2.

should be removed for showing. Campine cockerels with a lot of these cock-type saddle feathers should be culled.

Barred: The removal of odd black feathers from barred varieties has already been mentioned. For an explanation of why they continue to appear after over a century of careful breeding for the sharpest possible markings on every single feather, *refer* to the 'Barring' section under Double Mating, Chapter 12.

Cocks' Tails: The males of some breeds, including Cochins, Orpingtons, Pekins and Wyandottes, must have compact 'cushion' tails, with no large sickle feathers. They frequently grow feathers that project from the main body of plumage, and these are pulled out for shows.

Indian Game, Malay and Modern Game cocks are all required to have fairly modest length tails. Some have too much tail but are good in other respects, and in these cases a few key feathers can be pulled out about six to eight weeks before a show; by the big day, replacement feathers should have grown back to the desired length. Experts should be consulted for the exact timings for each strain.

SHOW PREPARATION

Washing Equipment

Most fanciers do their chicken washing in the kitchen sink – though it must be said that those who have a sink in a utility room or out-building will probably have fewer domestic arguments. If the kitchen is the only option, be sure to clear crockery and other break-ables before you start chicken washing. You will need:

- A bucket of warm water outside, to give muddy chickens' feet an initial wash;
- Bar(s) of soap;
- Shampoo (anti-dandruff or insecticidal for dogs) or washing-up liquid;
- Detergent/bleach powder (Ajax, Vim or similar);
- Hair conditioner;
- 'Glow-White' (for white plumage only);
- Roll of paper kitchen towel;
- Normal cloth towels: most fanciers use old ones, retired from human use;
- Plastic cocktail sticks;
- Old toothbrushes, nail brushes or similar, as required;
- Baby powder; this is only needed for white ear lobes and Spanish' faces;
- Baby oil and/or Vaseline for shanks/feet and combs/wattles (some fanciers use a special alcohol, camphor oil and glycerine mixture);
- Hair dryer;
- Clean boxes, baskets or (best) show cages placed near a radiator or other heater, to dry your birds in.

When to Wash?

Complete baths should be done at least three days before a show. Once done, and if the birds are kept clean, it should not be necessary to bath them again for several more shows. Many dark-coloured birds will need no more than a simple head and feet wash, which can be done the day before the show. In case they have more dirt than can be easily seen, on wing feathers perhaps, they should be checked several days before, so allowing time for a bath if necessary. Breeds that should be sleek and smooth-feathered should be bathed well ahead of the show: this is to allow time for the birds to preen and oil themselves, and so regain their smooth lines. Conversely, breeds that should have very profuse and fluffy plumage can be left until a little nearer show day.

Cleaning Feet

Start by simply holding the bird in one hand and washing the feet with the other. Novices might find it easier if they have a friend to help – though even experienced fanciers will appreciate some help when washing the largest heavy breeds. This initial wash is best done in a separate bucket of water so that the main wash water is not dirtied too much. Use soap, or Ajax or Vim, or any other product you think would be useful. I have never tried industrial hand-cleaning products, but they are certainly worth considering. Extra care is needed on feather-footed birds.

After an initial wash, lines of dirt can usually be seen under the edge of many of the scales, like dirty fingernails many times over. Very carefully, clean this dirt out using plastic cocktail sticks. This does not usually show on black feet, so is unnecessary. If this close examination of your chickens' feet reveals any cases of scaly leg-mite infestation, then treat the bird and do not take it to the show.

Scales are moulted every adult year at the same time as the feathers moult, though the old scales seldom actually fall off, but stay in place with a new scale underneath. Skilled poultry people know how to remove these old scales by gently sliding a cocktail stick between the old and new scales. New fanciers should ideally find an expert to show them how to do this as it could hurt birds if bungled or attempted on scales that are not 'ripe' for removal. These old scales are a slightly different colour from fresh, live scales.

Some breeds that should have 'clean shanks' (in this sense meaning not feather-legged) have a few rudimentary feathers growing out between the scales. These should be pulled out, and any visible holes hidden with a generous smear of petroleum jelly (for example, Vaseline). Accurate records should be kept of any birds so treated (another good reason for using PCGB numbered leg rings) because although they might win something at a show, they might not be good breeding stock. Without careful selection, a few minor feathers could lead to a lot of useless duds after a few generations. The breeds most likely to have this fault are those with Brahmas and Cochins in their ancestry, and include Rhode Island Reds, Sussex and Wyandottes.

The Actual Bath

It may not be necessary to completely immerse all birds, but it will be for any with white or other light-coloured plumage. Darker-coloured birds may only have visible dirt on a few places: wings, tail and around the vent perhaps, and in these cases, only the dirty parts need be wetted. If you are feeling nervous about all this, practise on another dirty chicken first, perhaps one not going to a show.

Start by half filling a sink with warm, not too hot, water. If a complete bath is intended, immerse the whole bird, except for its head. Then lift the bird out of the water and apply shampoo. Gently work the shampoo in, always from head to tail, in the direction the feathers grow. Use Ajax or Vim for any very dirty parts. Re-immerse to rinse, and repeat as necessary.

It was regarded as normal preparation until the 1960s to make white birds 'extra white' by standing them in a bucket of warm water, blued with 'Reckitt's Blue Bag', once a normal part of home laundry operations. Younger readers might need to ask their grandparents about

this product; the modern equivalent is 'Glow-White'. As before with Blue Bags, dissolve a sachet in water. Half a sachet in a large bucket will be enough for bantams, but a whole sachet in a larger tank or bath will be needed for large Cochins, Sussex or Wyandottes. As with the water, do not immerse their heads. This treatment gives a really bright white, and at the larger shows it is obvious that most established exhibitors use Glow-White, even more obvious under some types of fluorescent tube lighting.

After washing, and after Glow-Whiting where applicable, apply hair conditioner. Hold your soggy chicken for a few minutes for the conditioner to work, and then rinse off. This is very useful for smoothing down any poor quality plumage. Rhode Island Reds are prone to neck and back feathers that curl out slightly, and many of the hard-feather (Game) breeds have curly, sometimes called 'wiry', neck hackle feathers. A hair conditioner designed to improve 'dry and unmanageable' hair would seem most suitable.

Your bird now has to be dried, which can be a long process. Many fanciers do their bathing in the early evening so that the birds can spend all night in a box, basket or show cage next to a heater. It is best to start by gently wrapping the bird in a warm towel, taking care not to damage any plumage. As with the washing, only pass the towel from head to tail. Hair dryers are used most on breeds that should have a profuse and full covering of plumage, especially Silkies. Those that should have a tighter feathered, sleeker appearance are allowed to dry naturally, hence the need for a cage or large basket near a room heater. It is also why these breeds are ideally bathed a week before the show, to allow the feathers to flatten down properly.

Apply a little baby powder to white ear lobes and Spanish' faces the day before the show. At the same time, or even at the show as you put the birds in their cages, apply baby oil, petroleum jelly (such as Vaseline) or other 'secret mixture' to shanks, feet, combs, wattles and faces (except Spanish), taking care not to get any in their eyes.

TRANSPORTING YOUR BIRDS TO A SHOW

Everyone visiting a poultry show will have seen the range of containers used to transport birds to shows: there will be everything from cardboard boxes to expensive baskets and specially made wooden carrying boxes. One type of container you don't see a lot is the plastic commercial poultry crate, the sort intended for a dozen or so commercial layers or broilers. When used for show birds, it is usually ducks. There are two main reasons why these crates are not generally suitable for show chickens: first, they are not high enough for birds with large combs or tails; and second, having cleaned birds for a show, exhibitors do not want to risk birds fighting or attempting to mate during the journey to the show. Therefore birds are transported individually, or at most in twos or threes of birds who normally live together.

Plastic pet carriers, as normally used for dogs and cats, are very good. They are washable, which is a major advantage over traditional materials. If you are intending to keep some of the largest breeds, and quite a lot of them, the cost of the larger sizes of these carriers will probably be prohibitive. They are very useful, however, and if your local pet shop has a special offer, or you spot one going cheap at a car boot sale, then buy it.

Open, plastic-covered weldmesh pet containers are really most useful for drying birds after a bath, as described above. Their disadvantage is that tail feathers could poke through the holes in the mesh and be damaged, though this is obviously not a problem for those keeping compact-tailed breeds.

Traditional wicker baskets are lightweight, allow a free airflow, and are useful apart from being very difficult to wash. They are also expensive, so again, look out for cheap offers.

Some people look down on cardboard boxes, but these have several advantages: they are cheap or even usually free, and from the hygiene point of view, can be chucked away after use. Boxes will obviously need to be large and strong enough for the breeds to be carried, and a number of holes cut through the cardboard for the birds to breathe easily during their journey; in particular, ensure that enough holes are clear when several boxes are being loaded into your vehicle. Finding suitable boxes for the largest breeds might be more difficult. Tea chests are excellent if you can get them, otherwise look out for really heavy-duty cardboard boxes as used for domestic appliances.

All types of container should first have a paper feed sack or old newspaper placed at the bottom, and then a covering of wood shavings to absorb the moisture from droppings passed during the journey.

When putting birds in their cages at a show, take care that all their feathers, especially wing and tail feathers, are correctly in place. I also think they are happier, and settle in these strange surroundings better, if they are given a little food, though not all exhibitors agree. Novice exhibitors might think this a perfectly natural and humane thing to do, but it has been controversial in the past because it was used to cheat. Exhibitors would feed a visibly distinctive seed mix so that their friend who was judging that day would know which bird to make the champion. Suspicions of such malpractices are much less common now, and the public at summer agricultural shows often object if they see birds without food or water. It is therefore best to give just a little before judging, then your birds will have eaten it all before the judge gets to your bird. Old English Game and some other breeds should certainly not have more than the lightest snack, if anything, before judging because a full crop spoils the desired broad-breasted, short-backed body shape. It is best not to put water pots in show cages until after judging, although as said above, you may be obliged to do this before judging if there is going to be a lot of general public attending. If you are in doubt, check with the show secretary. Breeds with crests and beards are definitely not given water before judging, and the same applies to waterfowl because they will enthusiastically make a mess.

You have now done all you can, so stay away from your exhibits until judging is finished.

CHAPTER NINE

Judging Systems: Grading or Comparison

Grading card judging is the normal system used throughout mainland Europe, with comparison judging being the norm in Britain, Australia and the USA. Some British breed clubs have tried a mixed system for their annual shows, namely comparison judging with comment cards as an 'extra'. However, apart from such experiments, it is unlikely that the poultry showing community of any country will switch completely from one system to the other; the hobby has been shaped by the judging system and local circumstances to a much greater extent than many fanciers realize.

A German judge, for example, will have to fill in a detailed comment card for every one of the maximum of 100, usually nearer eighty, birds he or she is allotted. On the basis of these comments, each bird is given a grade. These are detailed below for several countries (*see* pages 79–80), but generally go from 'Excellent', via 'Good', to 'Inadequate'. German and Dutch poultry shows tend to be very large events and often last for three days, with the whole of the first day taken up with judging. There are really huge entries of the most popular varieties, and the more realistic breeders will go home perfectly happy if all their birds get fairly high grades. They might dream of winning 'Best of Breed', but really expect the top prize. Those who keep very rare varieties, as in Britain, exhibit to remind the fancy as a whole that their chosen favourites are not extinct yet. Rare breeds get better recognition

under the grading system, as some can still be awarded 'Excellent' if they are good enough. This is also true of rarer or more difficult colour varieties of otherwise popular breeds. With some breeds there is a tendency for one colour, often white, to win 'Best of Breed' every time. This might still happen in Germany, but at least the breeders of the other colours might get some high grades.

British judges only (only!?) have to award first, second and third places for each class. Some shows have fourth (or 'Reserve') places as well, and a few still follow the tradition of fifth ('VHC' for Very Highly Commended), sixth ('HC') and seventh ('C'). Judges are given judging forms with a tear-off results' strip for the show secretary; the comments they write on the part of the form they keep are only to remind them as they work through the judging, but if it is a breed club show they will be expected to produce a judge's report for the next club newsletter. While they are actually judging, all they will write is an odd word or even shorthand code; 's/w' for 'split wing', for example. Every bird will be taken out of its cage and examined, even though the judge has spotted at first glance that some of them have obvious serious faults and are not going to win anything; the owner of a dud bird might claim it wasn't given a fair chance if he/she saw the judge walk right past it. If it is a close thing, perhaps between second and third, a British judge may have to spend a lot more time deciding between them than

Famous British poultry judge Will Burdett considering this class-winning Silver Pencilled Wyandotte bantam for higher honours.

would their German counterpart, who would just give both birds the same grade.

Despite occasional difficult decisions, British judges can inspect at least 150 birds before lunch, and sort out the major awards in an hour after lunch; so at a local show here, everyone can take their birds home in one day. British exhibitors in the north of England are possibly more competitive than most southern English or European breeders, and once the prize cards are up they seem to want to go home: they will already be thinking of next week's show. From October through to February there are enough shows within reasonable travelling distance for a keen exhibitor to be out most weekends throughout the North and down to the Midlands. Breeders who live around the edges, the East Anglian coast for example, have fewer

shows, unless they are willing to drive a long way.

But the English definition of 'a long way' is nothing to the distances travelled by American and Australian exhibitors. Fanciers like to have as many different judges as possible, so to ring the changes clubs will arrange for judges to fly in from hundreds of miles away. They will have to be accommodated for a couple of nights, either at the home of one of the show committee, or at a local hotel. This can be ruinous for a local poultry club's finances, so comparison judging enables each judge to do the maximum number of birds – an American judge might have to assess 400 birds in a whole day of judging. Because a lot of the exhibitors will have travelled a fair distance as well, there will generally be a second day to socialize.

Although the prize cards may go down to seventh, many exhibitors are only interested in first or second places. British exhibitors seldom enter more than three birds in a class, because there is no point in doing so under our judging system. Rare-breed supporters may sometimes enter more to give a bigger display, especially at summer agricultural shows where there will be a lot of public to see the show. Those with some stock to sell might also enter them in a show, but these motives are not directly related to the competitive aspects of the show.

In all countries there are breeders with huge collections, and a majority with much more modest flocks. For any given total flock size, it is generally true that British breeders will have a higher percentage of different varieties, and therefore fewer of each type, and German breeders will specialize in fewer varieties, but obviously with a lot more of each. The different judging systems are a major factor in causing these flock differences.

As much of this book makes clear, specialization and larger flock sizes are needed to produce a succession of good quality show birds, which also conserves our breeds for the future. In these respects, many British breeders would benefit from specializing more than they do now. Nevertheless, there are some advantages to having a more diverse flock, in so far as those varieties that have less general appeal – some of the rarer colours of a breed, perhaps – may not be anyone's main speciality, but are still kept in small numbers as secondary varieties in larger mixed collections. Such large collections are often a very valuable source of stock to new fanciers, this being particularly true in more isolated areas such as Ireland and the north of Scotland, where a handful of breeders can be vital for the whole hobby.

EUROPEAN GRADING CARD JUDGING

All poultry shows in Germany and The Netherlands are large scale, rather formal events. They do not generally have British-style small local shows, although some breed clubs may hold specialized events that are more similar in general atmosphere. Grading card judging and smaller, more informal events are found in Belgium and France, however.

At the larger events, all the show cages are at a single level and this perhaps makes these shows look even bigger than they really are to British visitors. A string held up by canes runs about 20in (50cm) above the cages and the completed grading cards are clipped to it. Judges are expected to write suitable comments about each bird as well as just giving the grade, and this can be difficult to put into words in some cases, especially when the differences between the exhibits in a class are matters of body shape. Imagine trying to accurately describe in a few words the differences between each of fifty or more Black Wyandotte bantam pullets in a single class. Of course similar birds are given the same grade, but even so the comments have to vary a bit.

It may present the judges with some difficulties, but their problem is a positive advantage for the exhibitors in that the judges are forced to justify their decisions. As the judges know this is the case, they will be more inclined to swot up on breeds they may have to judge, but do not know much about. British judges are more inclined to muddle through on a combination of judging on general condition, general defects and guessing on any unknown breed details. This has been most true of some older hard-feather judges who have been less than keen to learn about the Asian breeds.

German Grades

V	= Vorzüglich	= Excellent*
Hv	= Hervorragend	= Distinguished
Sg	= Sehr gut	= Very good
G	= Gut	= Good
B	= Befriedigend	= Satisfactory
U	= Ungenügend	= Inadequate
OB	= Ohne Bewertung	= No assessment
NA	= Nicht anerkannt	= Not recognized (in the standard)

* (very few of these awarded) 'Blaues Band' may also be awarded to outstanding 'V' or 'Hv' exhibits.

Dutch Grades

U	= Uitmuntend	= Excellent
F	= Fraai	= Beautiful
ZG	= Zeer Goed	= Very good
G	= Goed	= Good
V	= Voldoende	= Sufficient
M	= Matig	= Moderate
O	= Onvoldoende	= Insufficient

Belgian Grades

AA
A
BB
B
C

Judges in each country soon adjust to the appropriate grade for exhibits of various qualities according to the general standards of the shows. Some 'AA' exhibits in Belgium will be as good as some 'V's in Germany, but certainly not all of them, as an 'AA' does not need to be quite as perfect. The knack for getting the right grade is similar to that of people who work in clothes shops who can usually estimate the correct size of customers at a glance. In similar vein, some birds, as some customers, may prove to be one grade/size higher or lower than was first thought on close examination.

After all the birds have been graded, the top awards are awarded. In Germany, the 'V' grades and 'Blaues Bandes' have to be confirmed by a second judge.

British shows have a 'schedule of classes' that is different for each show. Many breeds, or rarer colours of locally popular breeds, will be in mixed 'AOV' or 'AOC' classes. There are no such mixed classes in mainland Europe, no schedule of classes at all, in fact. Exhibitors just fill in all the bird details (breed, colour, sex, large/bantam), which are then entered by the organizers on a standard computerized show programme. Even though American shows are judged by the comparison system, they use the same entering system.

COMPARISON JUDGING AS IN THE UK

British Poultry Standards includes a 'scale of points' for each breed, but judges never, or virtually never, score points when judging. The points system has been there since the early editions in the nineteenth century, and no one has been brave enough to suggest abandoning them. They are of some use, however, as an indicator of the relative importance of characteristics on each breed.

Judges are given a judging book, or loose judge's forms, to record their results. There is a perforated strip on the right of the form from which the show administrators will fill in the prize cards. The judge will keep his/her book/forms, especially if a report is expected by a breed club.

As the judge initially approaches each class, any absentees are noted, as are the basic details of the birds in mixed classes. There will be a lot of more or less mixed classes, indeed at small local shows most of the classes will be mixed. British exhibitors like some competition. If there is a class for Large Leghorn, any colour, M/F, the judge will note 'Black M', 'Brown F' and so on for each bird. If there has been a much larger number entered than expected, the show secretary may have divided the sexes into two separate classes. As some of the exhibitors may not have fully completed their entry forms, it is sometimes left for the judge to separate the sexes when he/she sees what has actually arrived. Depending on the mix of Leghorns (in this example), the judge may instead choose to divide from the other colours whichever colour variety there is most of.

Each bird will be examined, both as they stand in the cage, and in the hand. The following chapter covers general defects that the judge will look out for, and there are also all the specific breed requirements to be assessed, as described in the *Standards* book. The judge will make notes, usually using some kind of personal shorthand, to keep track of the good, the not quite so good and the completely hopeless. When all the birds in the class have been examined, the judge will go back for a second look at any two or three birds of very similar quality, and will note those class winners that are possibilities for higher awards.

When all the classes are finished, the judges get together to find the section winners. Most shows now have a 'championship row' of cages for the section winners. The sections at British shows are:

Large Hard Feather	Bantam Hard Feather
Large Soft Feather*	Bantam Soft Feather, Heavy
Waterfowl	Bantam Soft Feather, Light
True Bantam	Rare Breed (Large and Bantam)
Trio	Juvenile's Exhibit
Eggs	Turkey

* At the National Show, Large Soft Feather is divided into Heavy and Light.

The breeds in each section are listed in *British Poultry Standards*. There are occasional changes, usually when a new breed club is formed for a rare breed that has attracted a lot more support.

If more than one judge has been involved in one section, they will go to the cages containing the possible section winners, to jointly agree on the 'Best'. Many shows include regional breed club shows that will have a specialist judge. In most cases each section will be entirely judged by one person. Once all the section winners have been decided, the next step is to find Show Champion and Reserve Champion.

This can sometimes be a long and discursive process, although thankfully the judges usually agree within a few minutes. Some show organizers, perhaps those with bitter memories from previous years, will appoint the most senior judge (in terms of judges' tests passed, not age) to choose the champions alone. If the championship judging is to be done jointly, it is important that all the judges, including specialists, have a reasonable all-round breed knowledge. If there is a long argument between the judges, organizers sometimes ask one of the exhibitors who is also a qualified judge to act as referee.

Comparison Judging and Breed Popularity

There is an optimum number of entries in a class for comparison judging to be successful, probably between eight and fifteen birds at most shows, up to twenty at a major event. Prize cards may be awarded down to seventh place, but most exhibitors are only interested in first, second and third. If there are at least eight birds from three exhibitors, there will be enough to interest the fanciers, and if there are twenty birds from (say) eight exhibitors, they will all have a lot to discuss while the judge is doing his duty. Most genuine fanciers are friends as well as rivals. If a variety is very popular, with thirty or more entries regularly appearing in a single class (for example Black OEG Bantam pullet), potential new breeders might be put off because they think that they have little chance of winning against established experts. They might decide to keep a less popular variety instead. Where the whole poultry fancy is of limited size, and many varieties are very rare, this is beneficial; however, it does place a limit on the potential for our hobby to grow to the size it is in Germany.

CHAPTER TEN

General Defects

Some of the defects described here apply to all, or nearly all, breeds, while others are applicable to a more limited range, but are nevertheless broader than specific breed faults. Many of these defects require automatic disqualification by judges at poultry shows; others are still faults, but not as serious. A bird with one of these faults might even win a third or fourth prize card if there are not many entries and some of the others have major breed faults. Ideally no birds with any of these faults will be used for breeding, certainly not cockerels, though in very rare varieties it may be necessary to keep some pullets with minor faults.

Side Sprigs on a Single Comb

These are extra fleshy points on the side of the rear part of a single comb. It is a fault that merits disqualification, although there is some scope for argument on combs with a slight bump on the side. How big does it need to be for disqualification? Side sprigs on very small-combed bantam pullets are sometimes missed by judges and their owners, and this can be an unfortunate mistake if these birds are bred from, and completely unshowable side-sprigged cockerels are bred from them.

Two breeds from Spain, Empordanesas and Penedesencas, are actually required to have side sprigs. Penedesencas lay very dark brown eggs and so seem likely to attract a lot more interest. There are (in 2004) a few breeders in the UK, but the birds have not been seen at the shows, or standardized. They will probably cause a lot of controversy among more conservative judges. This fault might be caused by the 'Duplex gene', which produces horned and cup combs, or another unknown allele at the same chromosome locus.

Double or 'Fish-Tail' Points on a Single Comb

This is a large comb serration that ends in two small points. Large single-combed cockerels such as Anconas, Leghorns and Minorcas with this fault can be killed as soon as they are spotted, as they are useless for breeding or

Side sprigs on a single comb.

Double or 'fish-tail' point on a single comb.

'Thumb marks' on a single comb.

showing. Double points are much less obvious on their floppy-combed sisters, which may be why the fault persists. Or maybe it is just a

Badly curved single comb.

matter of chance? The Dutch Bantam Club regards double points as a very serious fault because it is so regarded in The Netherlands; however, it is not regarded as quite so serious on large heavy breeds such as Orpingtons and Sussex. There have been quite a few champions of these and other heavy breeds over the last century which still won, despite their odd comb formations, because they were bigger or better coloured than their competitors.

'Thumb Marks' on a Single Comb

These are indentations on the lower part of a single comb. They are usually seen on males, and may be a result of the birds being kept in housing that is too warm. One of the functions of combs on chickens is to be a cooling organ. Domestic fowls are derived from jungle fowl species from tropical South-East Asia.

Curved Single Comb

This is where the rear part of a single comb curves round to one side. It may also be associated with combs growing too large.

Flyaway single comb.

Flyaway Single Comb

This is a usually smallish single comb where the lower edge of the rear part of the comb follows a rising line. This type of single comb is actually required on Nankin bantams, and is not required, but is often seen, on several European light breeds including Friesians. A flyaway comb is regarded as a more serious fault on breeds with larger combs.

Double-Folded Single Comb, and Floppy Single Comb Blocking Vision

These are both less than ideal formations on Leghorn, Minorca and other breeds with a floppy comb on the female. Neither is particularly serious, but if show success is to be guaranteed, it is necessary to breed as many pullets as possible to have a lot of choice.

Rose Comb that is Too High, or Over to One Side

This is usually a problem of males, as on most rose-combed breeds the females' combs are too small for this to occur. The possible exceptions are Derbyshire Redcaps, Assendelfters and

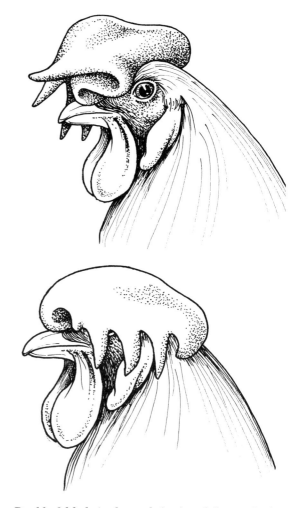

Double-folded single comb (top) and floppy single comb (bottom), both obscuring the vision.

some colour varieties of Barbu d'Anvers bantams. It is quite common for cockerels in their first adult year to have very attractive, neat, tidy combs, which by their second year have become very overgrown and poorly shaped. This is most likely to happen on White Wyandottes and rose-combed Anconas and Leghorns. Some cockerels of these varieties have combs that are rather small in their first year, but they will be the best older show birds.

Black Barbu d'Anvers males usually have very close-fitting combs, but some Cuckoo and Quail Barbu d'Anvers have very high combs.

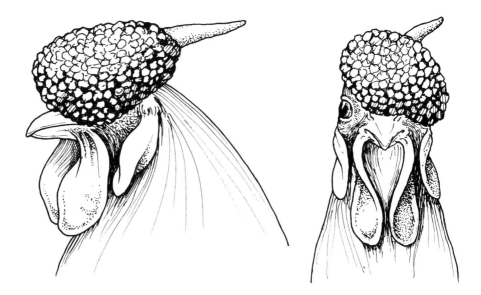

Too high (right), and lop-sided (far right) rose comb.

Rose Comb with a 'Hollow Front'

The front of a rose comb should be solid. Some have a dished or hollow front, almost resembling the front of two parallel single combs.

Rose Comb with the Leader at an Incorrect Angle

The leader of a rose comb is the point or spike at the rear, and it is clearly stated in each breed standard which is the correct angle. They range from leaders that follow the line of the skull, as on Barbu d'Anvers and Wyandottes, to leaders that sweep slightly upwards, as on Hamburghs and Rosecomb bantams. Leaders which point straight up are a definite fault on all breeds. These sharply angled leaders can be easily seen on very young growers, and any birds with combs like this should

Hollow-fronted rose comb.

Rose comb with the leader at an incorrect angle, either sticking straight up or sharply to one side.

85

never be used for breeding and are obviously useless for showing. Cockerels like this can be killed and pullets sold as pets/layers. The same is true of crooked leaders, and leaders with double points.

Rose Comb with an 'Ingrown' Leader or No Leader at all

Young birds with these faults should face a similar fate to the previous example. 'Ingrown' leaders look as if they have been pushed in. There is, however, a rare breed in The Netherlands, the Starumse Rondkam, that is required to a have large circular rose comb without a leader; however, this is the exception.

Rose Comb that is too Coarse or too Smooth on the Top Surface

The correct amount of 'workings' is also specified for each breed in the standards, though in each case the normal reality is sometimes slightly different from the written standard. Novice breeders should study closely as many birds as possible regarding this, and every other aspect of their new breed. Derbyshire Redcaps, Assendelfters and Starumse Rondkams have the largest 'workings' (papillae) as standard. Wyandottes, or at least some colour varieties of Wyandotte bantams, have the smoothest rose combs. It is from these varieties that most, or all of the strains bred in Britain have come from, originating from Germany or The Netherlands, notably Barreds, Buffs, Black Mottles and Salmons. White Wyandottes in Britain have no European connection, and usually have larger, less smooth combs.

Where there is some variation in comb formation in a strain, it is considered ideal to mate coarse-combed males with smooth-combed females, or vice versa.

Pea Comb that is too Tall and Narrow

Ideal pea combs consist of three lines of small papillae, running from front to back. The centre line should be slightly higher than the two outer lines, but not so high as to lose the general neat and compact form desired. Some pea-combed breeds have very high and narrow combs, even

Rose comb with an ingrown leader, or no leader at all.

to the extent of flopping over. Sumatras and pea-combed strains of Yokohamas are most likely to show this fault. Whilst no doubt partly an inherited fault, it is probably made worse by the fact that these breeds are often kept inside, in warm houses, to avoid dirtying their long tails.

Irregular Pea Comb

British-type Araucanas, the crested and tailed ones, often have very irregular-shaped combs, where the three lines of papillae are far from straight. This may partly be caused by the comb being somewhat compressed by the crest behind the comb, and partly by the persistence of walnut combs in this breed.

Poorly Divided Cup Comb

Cup combs are characteristic of Augsburgers and Sicilian Buttercups. This formation is caused by the 'Duplex gene' interacting with the single-comb gene, and so comprises a divided single comb. Ideally, only the front one or two serrations should be as one single comb, with the remainder clearly divided to give a large cup or crown effect.

The most common fault on these breeds is to have three or more points as one single comb, leaving a very small crown at the rear. In fact these rather poor combs are more common than good ones, so large numbers of chicks need to be hatched each year to be sure of getting some good combed show birds.

Uneven or Misshapen Wattles

Whether a breed is required to have small, neat wattles or long pendulous ones, in all cases both wattles on each bird should be the same size. Sometimes one wattle just never grows or is obviously compacted and deformed. This may be inherited, but is more likely to be the result of an injury or local infection. Major differences in size between wattles should be obvious, but novices may miss less dramatic examples; they should remember to check this point closely. Show judges will usually spot this immediately.

Clearly affected birds cannot be exhibited, but as this defect is probably not inherited, hens may be used for breeding if you are short of stock. Most breeders will have more than enough good cockerels, so those with defective wattles can be culled.

Wattles and Beards

Bearded breeds should ideally have minimal wattles. Noticeable wattles, large enough to detract from the beard, are seldom a problem on females, but some cockerels will grow wattles and these will usually be culled.

Incorrect Eye (Iris) Colour

Ideal eye colour is specified in the breed standards. With some breeds, a bird with the wrong eye colour would be instantly disqualified at a show, whilst with other breeds the judge may not even notice. This could happen with the same judge doing both classes at one show. Breeds in which correct eye colour is critical include all the hard-feather (game) breeds and breeds with large white ear lobes. Incorrect eye colour might go unnoticed on breeds with an intricate plumage pattern.

Eyes with Odd-Shaped Pupils or with One Obviously Blind Eye

Judges are required to 'pass' (disqualify) any bird with these defects. As may be imagined from the above, eye defects are likely to be noticed in some breeds more than others. Blind eyes will often appear cloudy. Blind eyes, or eyes with odd-shaped pupils, may be inherited, or be the result of an injury, or they might be a 'one off'. Affected hens might therefore be bred from if you are short of females.

Deformed Beaks

Crossed beaks and 'parrot beaks' are sometimes found among groups of chicks. These conditions may be caused by vitamin and/or mineral deficiencies, especially vitamin D and manganese. Kill all affected birds.

White in Red Ear Lobes

Positive ('enamel') white is a disqualifiable fault on red-lobed breeds, although judges do not always abide by this. Very pale red lobes, which may approach white, are often seen on birds that have been housed inside for some time. It is most likely to be seen on white or buff-plumaged varieties that the exhibitor was anxious to keep clean and out of the sun to avoid brassiness (on whites) or fading (on buffs).

Red in White Ear Lobes

Birds with perfect white lobes as they reach maturity for their first winter-show season often have reddened patches on their lobes by the next summer. This is partly just age, partly the sun and partly the effects of wind. Because this reddening is caused by age and their environment, it is perfectly safe to keep them as breeders. It has been well known for well over a century that some breeds have a very short show life, but actually live as long as any other breed. The most extreme examples are Minorcas and Rose-comb bantams.

White Extending from Lobes to Face and Eyelids

This is the other reason that Minorcas, Rose-combs and others have a short show career. As the ideal show bird has the largest possible lobes, and as the larger the lobe, the greater the chance of white spreading, this is bound to be a persistent problem. The fanciers are breeding to the limits.

Wiry Neck Feathers and Frizzled (Curled) Back Feathers

These problems are partly genetic, partly nutritional, and partly a result of general condition. It can occur on any breed, but is usually seen on Rhode Island Reds and the Asian hard-feather breeds. It can be minimized by ensuring the best possible diet (*see* Nutrition, Chapter 5) and by the application of hair conditioner (*see* the show preparation chapter, page 68).

Twisted Tail Feathers

This is not the same as wry tail (*see* below). In wry tail the whole tail is over to one side, but in this case most of the tail is in line with the body but just a few feathers grow out at an odd angle. This problem is most obvious on males, but can happen on females as well. I don't know how inheritable this problem is, but think it wise to assume it is an inherited fault; therefore it is safest not to breed from birds with this fault. It is clearly important to identify pullets with this fault, some of which can be hard to spot.

The easiest birds to see like this are cockerels of those breeds with large sickle feathers. More difficult are breeds with compact tails composed entirely of small feathers such as Cochins, Orpingtons, Pekin bantams and Wyandottes. On the plus side with these breeds, it is easy to pull out the offending feathers so they can be successfully exhibited. Nevertheless, records of leg ring numbers should be carefully retained so that these birds are not bred from. There is a high chance that if two such birds, each with just a few twisted feathers (easily pulled out), breed together, they may produce chicks with so many twisted feathers they will be beyond use.

Dented or Bent Breast Bones

Also called 'bent keel', this is caused by allowing young birds to use narrow perches. As most people will have seen on the commercial broiler chickens they have eaten, the breast bone of young chickens is composed of soft cartilage instead of hard bone. (*See* Housing and Equipment, Chapter 4, for appropriate young bird housing.)

Roach Back and Other Back and Hip Deformities

All affected birds should be killed as soon as they are identified. Some cases are easily spotted at about a month old, while it might be another month before birds that are less badly affected are noticed. Experienced breeders take as much time as they can spare to watch their birds. They will pick up growers for close examination, which also helps to tame them, as well as giving themselves the opportunity to spot these sometimes less than obvious faults.

Do not confuse these deformities with the characteristic curved back of the Malay breed.

Wry Tail

This is a condition related to roach back, but less serious. The tail is held permanently over to one side. All birds will hold their tail to one side sometimes, especially when in the confines of a show cage. A suspect bird should be closely studied for a few days, just to be sure, before killing it. If plucked for eating, it will then be seen that the parson's nose is bent round to one side.

Wry tail does not appear to affect birds in their ability to lead a normal life, but it is certainly a defect to be avoided. I don't know if it is inherited (or partially?), but it is better not to breed from any affected hens. You may know someone keen to give a 'good home' to such defective, but not actually suffering, animals.

Squirrel Tail

In these birds the tail is carried so high that the top edge of the tail feathers are carried forward of an imaginary vertical line; so the tail is in fact pointing towards the bird's head. Some breeds are required to have almost vertical tails, but forward of vertical is regarded as a fault. Japanese (Chabo) bantams are more controversial in this respect, and have been for a long time. Squirrel tail was included among the faults in the British standards of this breed, as it was for every other breed. Then in 1926 some recognized champion Japanese were imported to Britain and Germany from Japan, and the tails were pointing well forward on all of them. This caused a lot of serious debate here and in Germany at the time.

Distended Crop, and Drooping Abdomen
These are indications of a nutritional or other health problem, and birds with these problems should be disqualified.

In-Kneed or Knock-Kneed Legs
The legs actually turning at the hock joint is a sign of weakness, most often seen on longer-legged breeds. Sometimes birds that are knock-kneed when they just reach maturity will stand much straighter after a few more months. As experience is gained with any particular breed, it will become clearer which birds are likely to improve, and which are hopeless cases. Because all fowls have curved legs when seen from the front, it is important to note that it is the inside leg line that should be as straight as possible.

Bow Legs
Obviously, this is the opposite of in-kneed.

Stork-Legged
This term is applied to birds (usually to Old English Game) that stand too straight and stiff-legged.

Duck-Footed
This does not mean a chicken with webbed feet, although partially webbed toes do appear occasionally; it means that one or both hind toes are round the side of the foot, almost pointing forwards, instead of pointing back as they should.

Birds with this fault may lose balance. This was a (literally) fatal flaw on fighting cocks, and as Old English Game have been a very

popular breed at the shows since the 1850s, this has remained a disqualification for all breeds.

Crooked Toes
An obvious disqualification, although slightly curved toes might not be noticed. This may be inherited, or it may be caused by a slightly incorrect incubation temperature.

Bumble Foot
This is a swollen ball of the foot. It is most likely to be seen on heavy breeds and is often caused by injuries received when jumping down from perches that are too high.

Traces of Foot Feathering on Clean-Legged Breeds
Tiny feathers can be found down the outside of the shanks and between the toes, a feature most likely to be found on the more profusely plumaged breeds such as Orpingtons, Sussex and Wyandottes. These tiny feathers can be pulled out and the holes hidden with Vaseline for showing, a practice considered perfectly normal. If there is not too much unwanted fluff, and the bird is very good in every other respect, it will probably be worth breeding from. There will be others, however, that are too bad to breed from at all, and even too bad to be exhibited.

Adult Cocks with No Spurs
Some judges think this a significant sign of a cock bird being less than fully masculine and therefore possibly infertile, and will not award them prizes. Other judges may not even notice as they are concentrating more on specific breed points (such as plumage pattern).

Hens with Spurs
This abnormality is regarded as meriting disqualification in The Netherlands, but is ignored here in the UK. Indeed, in this country spurred Old English Game and Sumatra hens are regarded as being extra good.

CHAPTER ELEVEN

Some Poultry Genetics

A complete guide to the genetics of the domestic fowl would fill an entire book of this size, so this is necessarily a brief introduction. All serious breeders should closely study the subject, especially if they intend to cross colour varieties within a breed or revive an extinct variety. The author has met several fanciers who have made a lot of unnecessary work and expense for themselves because they lacked vital knowledge of breeds and the genes they carry.

This is not a static subject, so fanciers should obtain the most up-to-date information they can. Some of the genes confidently defined in older poultry genetics books – the two main works being Jull's *Poultry Breeding* (USA, *c.*1932) and Hutt's *Genetics of the Fowl* (USA, 1949) – have since been found to have been incorrect in several key respects. British breeder Dr Clive Carefoot of Preston, Lancashire discovered many of the true combinations of genes that give various plumage patterns, and described them in his 1985 book *Creative Poultry Breeding*. New work is likely to be published in poultry magazines and club yearbooks.

The first concept that novice poultry breeders should understand is that plumage patterns, and the colours in which the pattern appears, are determined by different genes. As domestic fowls are largely descended from the Red Jungle Fowl, their plumage pattern and colour is the logical starting point. This 'black-red'/partridge pattern, or 'Wild Type' as it is called by geneticists, can be bred in several colour versions. The 'black' parts of the

pattern can be changed to 'blue' (grey, really) to give 'blue-reds' or white to give 'piles'. Similarly, the 'red' or 'non-black' parts of the plumage pattern can be changed to white, as seen on 'silver duckwings', or an intermediate yellow shade on 'gold duckwings'.

THE E-LOCUS

This is the chromosome position where the most important plumage patterns are sited. There are probably eight genes that could be present at this locus, including the Wild Type pattern. Surprisingly, this gene is not the most dominant of this 'gene series'. They are, in probable order of dominance:

E or Extended Black: This gene adds to the black areas of the Wild Type pattern to give black plumage in all but the neck feathers of females and the neck and saddle feathers of males. These feathers have black centre striping, with the remaining parts of these feathers being 'non-black', which might be deep red, gold or silvery white according to the breed. This pattern is clearly seen on Norfolk Greys. Additional genes, which are not fully documented, give completely black-plumaged birds.

E^R or Birchen: As seen on Birchen and brown-red Modern Game. This is similar to extended black, but with breast lacing.

e^{Wh} or Dominant Wheaten: As on black-red/wheaten OEG.

e⁺ or Wild Type: As wild Red Jungle Fowl, Partridge Dutch bantams and many other breeds. The females have salmon breasts, striped neck feathers and 'peppered' back and wings.

eᵇ or Brown: This gene is usually found combining with other genes to give a number of patterns, as detailed below. It is probably seen in its least modified form on cock-breeder Partridge Wyandottes. The males are black-reds, virtually identical to the Wild Type gene above, but the females are different in that they do not have salmon breasts. This gene is the foundation of several other patterns in combination with other genes on different chromosome loci (positions).

eˢ or Speckled: Males are as Wild Type (e⁺), females are very coarsely stippled. This gene may be more important when it interacts with the pattern (Pg) gene.

eᵇᶜ or Buttercup: As its name suggests, this gene is seen on Sicilian Buttercups. It is modified by other genes to give the characteristic patterns of Pencilled Hamburghs, different for cockerel and pullet breeders (*see* Double Mating, Chapter 12) and several other breeds originating in Europe.

ABOVE: *Silver Sussex, large female. A variety showing the Eᴿ or Birchen pattern. This pattern is more commonly seen on Modern Game and Old English Game. When combined with gold (instead of silver, as here), this pattern is called 'brown-red'.*

Sicilian Buttercup female showing the eᵇᶜ plumage pattern. Its cup comb is acceptable, but not good. A better cup comb is shown on a 'Brocklesby Crown' bantam later in this chapter.

e^y or Recessive Wheaten: This is referred to in several sources, but Clive Carefoot proposed (*circa* 1980) that this does not exist as a separate allele, and suggested that some of the patterns and colour shades associated with this gene are in fact caused by other gene combinations. If the matter has been settled, the author has not seen the relevant material.

THE PENCILLING OR PATTERN GENE: Pg

The true importance of this gene has been highlighted by Clive Carefoot. The older poultry genetics books (Hutt, Jull) suggested separate genes for spangling (as Spangled Hamburghs), lacing, 'autosomal barring' (as Pencilled Hamburghs and Campines), double lacing (as Barnevelders and Indian Game) where they suggested genes at all. Clive's work suggested that a single gene 'Pg' interacts with the different genes of the 'E series' and others to give these patterns. Much of Clive's interest in these genes arose from his attempts to fully understand the genetic differences between cock-breeder and pullet-breeder strains of Partridge and Silver Pencilled Wyandottes. (*See* Chapter 12 for more details.)

This gene mostly affects female plumage patterns. A 'Wild Type' female, as seen in its most refined form on Brown Leghorns and Gold Partridge Dutch bantams, have fine black random peppering on their (brown) back and wing feathers. Genetically they are e⁺, pg⁺. The author has not had any direct experience of breeding Brown Leghorns, but does know that many partridge-patterned (in all colour combinations) Dutch females show more or less distinct patterning (as would be correct for Partridge Wyandottes) on their backs. This would suggest that they are genetically e⁺,Pg. Although this effect is a serious fault on Dutch and Leghorns, there are other black-red/partridge breeds in which the female plumage pattern is not as important when judging Game and long-tailed breeds for example, where more or less patterned back feathers regularly appear.

The really interesting interactions can be seen when either 'Pg' or 'pg⁺' is present with the brown gene 'e^b'. For example, cock-breeder Partridge Wyandottes are e^b, pg⁺, whereas pullet-breeders are e^b,Pg. This shows the need for double mating (*see* Chapter 12) in this breed. It is obviously impossible to breed perfect exhibition birds of both sexes from one strain if they genetically different. Additional patterns are obtained when the melantoic, Columbian, another pigment-restricting gene called 'Db' and the mahogany gene 'Mh' interact.

THE COLUMBIAN GENE: Co

This restricts black pigmentation to neck striping, wing markings and tail. It is seen on Light Brahmas, Light Sussex and Columbian Wyandottes in silver form, and Buff Sussex, Red Sussex and Rhode Island Reds in gold form. The Columbian gene also reacts with genes on the E locus. Co,e^Wh is suggested by Carefoot for Light Sussex (they have white undercolour) and Co,e^b for Light Brahmas (normally grey undercolour).

MELANTOIC: Ml (NORMAL MELANIN = ml)

As the name implies (melanin = black pigment), this gene darkens or adds additional pigmented areas to whichever pattern the other genes present would produce. Clive Carefoot suggests the following genetic combinations for these patterns:

e^b,Ml,Pg	= double laced as Barnevelders
e^b,Ml,pg⁺	= self black with yellow shanks/feet, as Black Wyandottes
e^b,Ml,Pg,Co	= laced Wyandottes (variety according to gold/silver, black/blue & 'I' genes)
e^b,Ml,pg⁺,Co	= Quail pattern, as Barbu d'Anvers

Every breeder of Quail Barbu d'Anvers will realize the relationship between the quail and Columbian patterns, as numerous 'in-between' patterned birds appear in these strains. There are also colour differences between strains, for example Belgian and British Quail Barbu d'Anvers have lighter breast colour than German and Dutch strains. It may be that some strains are based on e^b and others on e$^+$.

THE BARRING GENE: B

This is a sex-linked gene that stops and starts pigment production as the feathers grow to give the well-known barring pattern as seen on Barred Wyandottes and Scots Greys. The extremely narrow and sharp barring seen on Barred Plymouth Rocks is achieved by the presence of a gene ('K') for very slow feather growth, which allows for a lot of on/off sequences in the time it takes for a feather to grow. The same barring gene ('B') when on a rapid feathering breed gives wide, coarse, fuzzy 'cuckoo barring' as seen on Cuckoo Marans, Cuckoo Pekins and others.

The barring gene also interacts with the genes on the E locus, all of the above examples being based on E plus melantoics. Thus all these barred and cuckoo breeds would be self-black if they did not have the barring gene. This gene has a greater pigment-restricting effect on black pigment than it does on red or gold pigment. When the barring gene is applied to the Wild Type pattern, the 'Crele' variety is produced; and the barring on Columbian pattern combination is seen on Rhodebars, one of the autosexing breeds.

It is to be hoped that most, preferably all, readers of this book are already familiar with the principles of sex-linkage. They should be aware that the barring gene is not completely dominant to non-barring, which is why both dark- and light-barred males are seen, but only dark-barred females (*see* barring in Chapter 12).

Barred Plymouth Rock bantam female, UK type. Sharp narrow barring is obtained because this breed carries the slow feather-growth gene.

Cuckoo Marans, large female. Broad fuzzy barring is obtained by the same barring gene as on the Plymouth Rock, but in this case it is modified by the rapid feather-growth gene.

In most breeds with a Crele variety, Crele OEG for example, the richer coloured males that have only one barring gene are exhibited, but in the autosexing breeds only light males carrying two barring genes should be exhibited and (in normal circumstances) bred from. The phrase 'in normal circumstances' is used here because some inbred strains of autosexing breeds are sometimes revived by crossing with the other related breed, for example Brown Leghorns for Gold Legbars. Dark Crele males may be used as part of this process.

GENES FOR WHITE PLUMAGE: S, c AND I

White plumage can be caused by three genes: S, c and I. S is the sex-linked gene silver, as seen on Light Sussex. The opposite gene to it is gold (s). This gene only changes the 'red' or 'gold' parts of an E locus plumage pattern, not the black parts of those patterns.

Those readers old enough to remember the famous Rhode Island Red × Light Sussex sex-linked crosses that were once the mainstay of poultry farms before the age of hybrids will know that the silver gene is basically, but not completely dominant to gold, and that there are other genes affecting the shade of 'golds'. The crossbred cockerels are essentially the same colour as their light Sussex mothers, but have some gold on their shoulders, and the crossbred pullets are gold, but a few shades lighter than their Rhode father.

Recessive white ('c') is the gene responsible for most white breeds of fowl. It is a mutation that simply does not function as does the normal dominant gene (C) in making essential chemical components of pigments. Birds that are heterozygous for this gene (Cc) can make black pigment perfectly well, but are a little less than fully efficient at gold/red pigments. This is seen in British standard cock-breeder Partridge Wyandottes that have yellow, rather than rich orange neck hackles; these strains produce occasional white chicks.

The dominant white ('I') gene is mainly found in White Leghorns. It is not completely dominant, being less efficient at suppressing gold/red pigments than black pigment. In its heterozygous form, and depending which other pattern genes are present, it produces some attractive varieties, such as pile (as on Leghorns and most Game breeds), Buff Laced Polands and the best strains of Buff Laced Wyandottes (*see* double mating chapter for further details of breeding the pile variety).

Barring Gene in White Varieties

Some winning strains of white varieties carry the barring gene. Because the barring gene acts as a partial restrictor of pigment it is useful as a 'back-up' gene to give really pure, 'snow-white' whites. Without the barring gene, whites can be 'sappy' and very inclined to go brassy in the sun.

THE BLUE GENE: Bl

This incompletely dominant gene at first confused, and then fascinated, poultry breeders and the first generation of scientists who studied the then very new field of genetics. When blue birds are mated together, Blue Andalusians being the most famous blue breed, they produce 50 per cent blue chicks, 25 per cent black chicks and 25 per cent 'splash' chicks; when fully grown the latter are mainly white, with some random blue and black areas. But if Black Andalusians are mated with Splashes (neither are standardized for showing), the result is 100 per cent blue chicks. The blue gene normally gives a darker, almost black edging to most feathers; this is seen most clearly on the breast. Generations of careful selection have developed this naturally rather fuzzy edging into sharp, clear lacing. However, this lacing can easily degenerate, which is why many Andalusian breeders prefer to use the blue × blue mating (so they can see the quality of the parents' lacing) rather than the apparently more productive black × splash mating, where good or bad lacing genes may be present, but are unseen.

The blue gene only affects black plumage areas on other plumage patterns, and can be seen on Blue Laced Wyandottes, Blue-Red/

ABOVE: *Andalusian bantam male. From a Blue × Blue mating, only 50 per cent of the resulting chicks will be blue, and very few of these will have sharp lacing, even ground colour, good comb and lobes and be of the desired size. Although good specimens are very beautiful, this breed remains rare because few fanciers are willing to rear large numbers of birds for the few really good ones.*

Splash Carlisle-type Old English Game, large male. Probably the result of a Blue-red × Blue-red mating. OEG are mainly judged on body shape, so such 'off colours' can be successfully exhibited at UK shows. Such 'Splashes' would not be acceptable show birds when hatched by breeders of Blue-red Dutch bantams or Blue-red Leghorns, but they might usefully be retained for breeding.

Silver Spangled Hamburgh bantam female.

BELOW: *Black Mottle Barbu d'Uccles pair. Only Anconas can be relied upon to produce a whole batch of uniformly spotted youngsters. Other breeds with this pattern, like these Barbu d'Uccles, are usually very variable. Breeders of Black Mottle Barbu d'Uccles, Japanese or Wyandottes will be lucky if they breed a uniform, showable exhibition trio.*

TOP LEFT: *1. Gold Pencilled Hamburgh bantam, exhibition male.*

TOP RIGHT: *2. Gold Pencilled Hamburgh bantam, exhibition female.*

ABOVE: *3. Double Laced Barnevelders, large pair. British Standard with laced breast on male. Solid black breasts are required on Barnevelder males in Germany and The Netherlands.*

4. Gold Laced Wyandottes, a pullet-breeder pair. Note the dark ground colour.

5. Gold Laced Wyandotte bantam, an exhibition male from a cock-breeder strain. Note the lighter ground colour. The leader (rear spike) of the comb should be lower, following the skull.

6. Silver Laced Wyandotte bantam, exhibition male. Note the clear white shoulder/wing bow.

8. Blue Laced Wyandotte bantam female. Ideally the neck feathers should have more of the red-brown colour, less of the blue.

7. Silver Laced Wyandotte bantam, exhibition female.

ABOVE: *9. Black Leghorn bantam, exhibition female. Note the tightly 'whipped' (i.e. not fanned) tail, correct for British standards, and the bright yellow shanks and feet.*

ABOVE RIGHT: *10. Blue Leghorn bantam female. The excellent even colour of this bird is only obtained on a small percentage of each group hatched, and it needs a shaded run to avoid fading.*

11. Chamois (Buff Laced) Frizzled Poland bantam male.

BELOW: *12. Frizzle bantam females, Buff, Blue and Black-mottled.*

13. *Millefleur Barbu d'Uccles bantam female.*

14. *Porcelaine Barbu d'Uccles bantam female. The millefleur pattern is diluted by the lavender gene. It is quite normal to cross these two colour varieties of Barbu d'Uccles.*

15. *Buff Orpington bantam female, a show champion. It is almost impossible to achieve a perfectly even shade all over, and down each feather to the skin.*

16. *Quail Barbu d'Anvers bantam female. This is a 'normal quail'; there are also blue quail, lavender quail and silver quail variants.*

17. Pile Modern Game bantam trio. A supreme exhibit from a skilled fancier, David Young.

19. Pile Old English Game bantam male, as shown in the UK, with reduced tail.

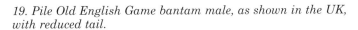

ABOVE: 18. Black-red/Partridge Modern Game bantam female, trained to pose.

BELOW: 20. Pile Old English Game bantam male, as shown in the USA, with large fanned tail.

21. *Blue Partridge Dutch bantam male. This variety is called 'Blue-red' by Game fanciers.*

22. *Black-red Yokohoma male. This bird would only be allowed outside in good weather.*

BELOW LEFT: 23. *Crele Old English Game bantam male, US type. The clarity of barring, especially on the tail feathers, makes this an outstanding example of this plumage pattern.*

BELOW: 24. *Crele Old English Game bantam female, US type.*

25. *Wheaten Old English bantam female, UK type.*

26. *Blue-tailed Wheaten Old English Game bantam female, US type.*

27. *Partridge Wyandotte bantam male, US standard: note the dark colour.*

28. *Partridge Wyandotte bantam female, US standard: note the dark colour.*

29. *Partridge Wyandotte bantam, exhibition male from a UK cock-breeder strain. Note the lemon top colour and solid black breast.*

30. *Partridge Wyandotte bantam, a female from a UK cock-breeder strain. Note the markings are less well defined than the exhibition female, and the lemon shade on the head and neck to match the males.*

31. *Partridge Wyandotte bantam, a male from a UK pullet-breeder strain. Note the more orange top colour shade and the spotted breast.*

32. *Partridge Wyandotte bantam, an exhibition female from a UK pullet-breeder strain. Note the perfect clarity of the markings.*

Partridge OEG, Blue-Red/Wheaten OEG and many other blue-based Game Fowl varieties. In Wyandottes, the 'splash' are blue-bred Buff Laced, which are often exhibited successfully at smaller shows, but are not as good as Buff Laced Wyandottes that have the white lacing caused by the dominant white ('I') gene.

Fanciers can have two recognized show varieties in one strain by keeping the normal (black-based) and the blue version of a breed as they are routinely mated together. Black × Blue (including patterns) gives 50 per cent of each. Old English Game breeders do this with the varieties named above. Normal Quail and Blue Quail Barbu d'Anvers bantam, and Dark and Blue Indian Game are examples.

THE LAVENDER GENE: lav

This is a simple recessive gene that dilutes black to pale grey ('lavender') and red to buff or cream. It can operate on any base pattern, changing black birds to self lavenders, as seen on Lavender Araucanas; and on the spotted millefleur pattern to produce the pretty and popular Porcelaine Barbu d'Uccles bantams.

SPANGLING

This pattern – large black spots at the end of each feather – is most developed on Spangled Hamburghs. Some other spangled breeds have much smaller spots, such as Appenzeller Spitzhaubens; and on Brabanters, Derbyshire Redcaps and Old English Pheasant Fowl, the spangling partially approaches lacing. The Spangling gene is probably also responsible, in conjunction with other genes, for the 'millefleur' pattern seen on Booted (Sabelpoot) bantams, Barbu d'Uccles bantams and Speckled Sussex (*see* Mottling, below).

Not all versions of spangling are fully understood genetically, or if they are it must have been discovered recently, and not widely published. As the simplest solution is most likely to be correct, it is best to look at the nearest related breed that is understood to suggest a mix of genes for each unknown version of spangling.

For example, Spangled Hamburghs, Derbyshire Redcaps and Old English Pheasant Fowl are all more or less related to Pencilled Hamburghs, Assendelftse Hoenders and others. Speckled Sussex are closely related to Red Sussex.

THE MOTTLING GENE: mo

This gene gives white tips to otherwise coloured feathers. Anconas are the most perfectly marked breed with this gene. It also changes Black-Red/Partridge OEG to Spangled OEG on the Wild Type pattern. The spotted pattern as seen on Speckled Sussex, Millefleur and Porcelaine Barbu d'Uccles and others is slightly more complex, as there may be different combinations of base pattern genes (brown, e^b and Columbian, Co, others?).

THE PIED GENE: pi

Exchequer Leghorns were the only breed with this random mixture of black and white known to European and American breeders for many years, but this plumage variety is now also seen on Japanese Shamo, Ko-Shamo and others. In Japan it is called 'Go-Stone', after the bowl of black and white counters used in the popular Japanese board game 'Go'. Leghorn breeders are keen to maintain the random mixture effect to prevent any confusion with Anconas.

BUFF PLUMAGE

This apparently simple self colour is very complex genetically, partly because the exact shade of colour seems to be different in each breed. They are likely to be based on wheaten (e^{Wh}), Columbian (Co), dark brown (Db), and two other colour-reducing genes, champagne blonde (Cb) and dilute (Di). British Standard Buff Plymouth Rock bantams are a very light lemon shade, caused by a single recessive white (c) gene. Buff varieties are very difficult to breed because they tend to be either too light (with white under-colour and/or white

Exchequer Leghorn, large female. This colour variety first appeared (in a flock of Blacks) in 1904 on Mr Robert Miller's Scottish farm. They were a popular commercial laying breed until all pure breeds were replaced by hybrids in the 1950s. Now they are a very rare variety in urgent need of more support.

peppering) or too dark (with bronze or black, usually in wings or tail).

RATE OF FEATHERING: K^n, K^s, K, k^+

This is a series of sex-linked genes that determine speed of feather growth. The genes are, in order of dominance:

K^n: extremely slow-feathering (Barred Plymouth Rocks?)
K^s: extremely slow-feathering (Barred Plymouth Rocks?)
K: normal slow-feathering, as on many breeds
k^+: rapid-feathering, Leghorns and others

There is a detectable difference in quick- and slow-feathering day-old chicks, the quick-feathering chicks having a little visible feather growth on their wings. It has been used, but not widely, for commercial sex-linked cross-breeding.

SHANK AND SKIN COLOUR

Apart from the Silkie with its very dark blue skin, poultry have either yellow or white skin. A slightly larger range of shank colours is available, including blue/slate, green/willow and black. There are two significant layers of tissue in the shanks: the dermis or under layer, and the epidermis or outer layer. Blue and green shanks have black pigment in the deeper cells, which gives these lighter shades because the surface scale tissue is (respectively) white or yellow.

The shade of yellow in skin and shanks is affected by diet and egg laying. Pigment builds up from grass and maize in skin and fat tissues, and is lost in egg yolks. If genetically yellow skin/shank birds are fed on a diet lacking these ingredients, their shanks will fade to a creamy shade, nearly white on laying hens. The author remembers a fancier, well known for his sense of humour, who had a single combed chick in a batch of White Wyandotte bantams. He won several prizes with it as a White Orpington bantam hen, much to the amusement of those of us who knew the true pedigree of the bird, and the embarrassment of the unfortunate judges when they were told later in the day.

Shank colour is also affected by some of the genes for plumage colour. Black shanks are only seen on birds with mostly black plumage, and the mottling and barring genes reduce black pigmentation to some spots or shading over a yellow (as Anconas) or white (as Cuckoo Marans) ground colour.

White shanks are caused by a gene ('W') which is dominant to its allele for yellow ('w').

The genes for dermal (under layer) melanin are sex-linked. The dominant gene ('Id') inhibits melanin production, the recessive gene ('id') allows it. This has caused some problems for Modern Game breeders who routinely cross Black-Reds (willow shanks) with Piles (yellow shanks) to maintain correct coloured Piles. Willow-legged chickens have yellow shanks as day-old chicks, the colour only developing when they are about six to eight weeks old.

Black-plumaged, black-shanked breeds fade in shank colour in their second year of age, Australorps to dark slate, Jersey Giants to olive green.

EYE COLOUR

Eye colour can range from nearly white ('pearl') to black, with yellow, orange, red and brown in between. Pink-eyed albino strains have been maintained by scientific institutions, but do not exist among standard-bred exhibition varieties.

Pearl and yellow eyes are important breed characteristics of the Asian Game breeds – Asil, Malay, Shamo – and are caused by a recessive gene. Chicks of these breeds have darker eyes, the final eye colour sometimes not stabilizing until the birds are over a year old.

Brown eye ('br') is caused by a sex-linked recessive gene, its dominant allele ('Br+') being for an orange colour, the shade depending on plumage colour.

Black eyes are associated with black plumage and skin colour. It is a permanent problem for Australorp and Black Orpington breeders, who

Close-up of a Spangled Asil male showing the desired very light eye colour. This colour develops with age, being more yellowish on young birds. Note also the compact pea comb and naturally minimal/absent wattles.

Close-up of an Australorp bantam male showing the desired dark eye colour. This bird also has a perfectly serrated single comb, essential for success among the large numbers entered at the major shows.

Excellent cup comb on a Brocklesby Crown bantam. This new breed was never standardized. If its creator, Mr Brocklesby, had instead made Sicilian Buttercup bantams, these would have been recognized without argument.

Close-up of a Silver Spangled Appenzeller Spitzhauben male, showing horned comb and pointed crest.

aim for black eyes with red combs and faces. They are frequently successful, but every group of youngsters will include birds that fail by having lighter coloured eyes or dark-pigmented facial skin.

GENETICS OF COMB TYPES

There are four main types of comb: walnut, pea, rose and single, which are determined by two pairs of genes. Pea ('P') is dominant to single ('p') on one chromosome position, and rose ('R') is dominant to single ('r') on another chromosome. If dominant genes are present ('PR'), the result is a walnut comb.

There is a link between rose combs and reduced sperm viability. Males that are homozygous for rose comb (R,R) have this problem, whereas rose-combed cocks carrying the single-comb gene (R,r) have normal fertility. For this reason the single-comb gene tends to persist in strains of rose-combed breeds, not visible until some single-combed chicks hatch.

Horned combs, as on La Fleche, and cup combs, as on Sicilian Buttercups, are caused by the 'Duplex comb gene'. There may be different genes, D^v for horned and D^c for cupped, or the two effects may be interactions between Duplex with rose for horned combs, and Duplex with single for cup combs. It might also be the case that the virtually combless Breda fowl has pea + rose (= walnut) + Duplex genes.

Comb formation is modified (pushed forward) by feathered head crests. Thus the 'Butterfly' comb of Houdans is a pushed-forward cup comb.

OTHER GENETIC VARIETIES

Crests: Cr

Crests are generally caused by an incompletely dominant autosomal gene: Cr. The author was not able to find any explanation of the differences between large, spherical Poland-type crests, pointed Appenzeller Spitzhauben and Brabanter crests, and modest 'Tassels' as seen on Tasselled OEG, Crested Legbars and Sulmtalers.

ABOVE: *White Poland bantam trio showing crests and beards. A magnificent exhibition trio that could only be produced by a specialist breeder, and an expert in bathing and other preparation techniques. A fancier with a huge menagerie of different varieties is unlikely to have the time to stage such a perfect trio.*

Silver Sebright bantam male. A henny-feathered breed. The eyes should be as dark as possible, as also should be the skin colour of comb, face, lobes and wattles. On Sebright males, an area of dark-pigmented skin around the eyes is all that can usually be obtained.

Henny Feathering: Hf

An autosomal incomplete dominant gene causes female-type plumage structure and (if applicable) plumage pattern and colours on males. It works by affecting the hormones in the feather follicles, and is a breed characteristic of Campines, Henny Game, pullet-breeder Pencilled Hamburghs and Sebrights.

Frizzled Plumage: F

An autosomal incomplete dominant gene causes curled feathers as seen on the frizzle breed,
as in Frizzled Japanese (Chabo) bantams, Frizzled Polands and Lyonnaise, a fairly new breed
from France not yet (in 2004) standardized in
the UK. Heterozygous (Ff) birds and normal
flat-feathered (ff) birds from these strains are
useful for breeding as repeated generations of
frizzle to frizzle matings eventually produces
rather pathetic-looking, sparsely feathered
birds.

Silkiness: h

An autosomal recessive gene produces Silkie
feathering, a fact well known to the many poultry keepers who have crossed Silkies with other
breeds to produce broodies.

Naked Neck: Na

An autosomal incomplete dominant gene causes the naked neck effect, which reduces the
number of feather follicles all over the body as
well as the neck. Heterozygous birds (Na,na)

have large 'ruffs' or 'collars' of feathers on the
front of their neck. There are now naked-necked
broilers, developed for the table poultry industry in warm countries. Breeders of exhibition
Naked Necks are likely to try crosses with these
to increase body size and reduce inbreeding.
Several generations of breeding may be necessary to achieve the desired result.

Muffs and Beard: Mb

An autosomal incomplete dominant gene causes this muffs and beard effect, which also
reduces the size of the wattles.

Ear Tufts

These ear tufts, typical of Rumpless Araucanas,
are caused by a semi-lethal gene. If birds with
good tufts are bred together there will be a lot
of dead embryos with head deformities.

Creeper: Cr

An incomplete dominant gene causing short
legs: (cr,cr) are normal leg length, (Cr,cr) are
short-legged, and (Cr,Cr) die as embryos during

White Silkie females, large and bantam. Apart from the dark blue feet and faces that can be seen in the photo, all the skin, and even the flesh is dark blue. 'Black Chicken Soup' is considered a delicacy in China and Japan.

Black Transylvanian Naked Neck bantam female. These need regular access to outside runs to keep the neck skin bright red; if kept inside for extended periods the neck fades to an anaemic-looking pale pink.

BELOW: *Black-Tailed White Japanese (Chabo) bantam male. An ultra short-legged breed with the Creeper gene. Some fanciers obtain better fertility by using slightly longer-legged males with very short-legged females for breeding.*

Multiple spurs on the shanks of a Sumatra male.

incubation. Typical of Japanese (Chabo) bantams, Scots Dumpies and similar breeds in other countries.

Multiple Spurs: M

An autosomal dominant gene causing more than one spur on each shank. The number may vary, but it is usually three, one larger central spur, with smaller spurs above and below.

103

Rumplessness: Rp

An autosomal dominant gene causing a lack of tail. Typical of Rumpless Araucanas, Rumpless Game, Barbu d'Grubbe and others.

Polydactyly (extra toes): Po, Po^d, po

Po causes the extra rear toe typical of Dorkings, Faverolles, Houdans, Lincolnshire Buffs, Silkies and Sultans. The other allele, Po^d, causes even more toes, and is most likely to occur on Silkies.

PLUMAGE FORMATION

There are certain to be many genes for plumage quantity and formation, length and shape of tail and so on, but these have not been well documented. When fanciers cross breeds to create new varieties they should be aware that it may be more difficult to obtain the desired type than to get the required colour. For example, if it was intended to create a new colour variety in one of the breeds with short but broad tails (Pekins, Wyandottes, and so on) it would be best to first look at the other available breeds with similar tails. It will be cheaper in the long run to obtain a similar shaped breed with the required colour genes, even if they will be expensive to obtain (for instance, if the nearest flock is a very long way away) than to use a readily available breed of completely different type. With some extreme types, any cross will be a very long-term project,

possibly longer than the life of the fancier! The author suggests that even crossing with other hard-feather breeds will never give acceptable new colours of Ko Shamo or Yamato Gunkei. Other fanciers of these breeds will certainly not appreciate any 'mucking about' with their historic favourites, anyway.

ABOVE: *Cuckoo Barbu d'Grubbe bantam male showing rose comb, muffs and beard, and rumplessness.*

Extra hind toes, showing the effect of the Polydactyly gene.

CHAPTER TWELVE

Double Mating

Scientific understanding of the processes of inheritance did not begin until the first paper by Gregor Mendel was published in 1866, and this was largely ignored at the time. For practical purposes, the science of genetics began with William Bateson, Britain's first Professor of Genetics (he invented the word) at Cambridge (1908–10); he had translated Mendel's work in 1900. Although Bateson did a lot of work on the inheritance of comb types and plumage colour of poultry, it was many years before this became common knowledge among the majority of poultry keepers.

During the same period, roughly 1860 to 1914, many of the breed standards were drawn up, and those of the ancient breeds refined. Even if the august gentlemen who ran the breed clubs had been closely following Bateson's work, which was very unlikely, the detailed understanding of the inheritance of plumage pattern and other points was unknown at the time. Thus several breed standards described a perfect male and a perfect female, which seemed achievable at the time, but in reality were a few genes different from each other.

Back in the early nineteenth century, a century before anyone could have said why, practical breeders of Lancashire Moony and Yorkshire Pheasant Fowls, later renamed as Spangled Hamburghs, knew that separate strains were needed to produce well marked males and females for their early competitions in local pubs. These separate 'cock-breeder' and 'pullet-breeder' strains were adopted for several more breeds between 1870 and 1890

when the poultry shows had started to become really competitive and breeders had to be able to produce near-perfect birds to win anything.

Most, but not all, of the breeds that have to be double mated have complicated plumage patterns in which the males are naturally a different pattern to the females. The partridge/pencilled pattern as seen on Cochins and British Standard Wyandottes is a perfect example. The markings and pattern of American Partridge Wyandottes are slightly different, designed to avoid the need for double mating, which did not suit American poultry breeders. However, the exact (very exact!) markings required to win at British shows needed two strains. This began by accident, with some breeders regularly winning with cocks, others with hens. The breed standards could have been changed to solve this, and get back to single pen breeding, but those with winning strains – and there was a lot of prize money at stake here – were very keen on keeping to the status quo. Eventually all this was explained in the major poultry books, but for a few years the existence of these separate cock and pullet strains was a trade secret. Champion exhibition trios could be sold for high prices with the vendor being secure in the knowledge that whoever bought the trio would not breed any serious competition for next year's show from the three exhibition birds. Some of these breeders were less than pleased when their secrets were fully described in the books.

There have been occasional complaints about the need for double mating ever since. Sometimes this has come from confused beginners,

White Leghorn bantam male, UK type. Note the large upright comb.

White Leghorn bantam female, UK type. Note the large floppy comb.

at other times from utility poultry breeders who either did not understand the background to the system, or chose not to. Such arguments seem to have been more in evidence in Germany, where some breed standards have been changed to make it unnecessary. In the case of Laced Wyandottes the standard was changed to accept pullet-breeder males as the exhibition norm. With Partridge and Pencilled Wyandottes there was a more radical change, and they were divided into separate varieties. Show standards were then drawn up for the previously unshowable cock-breeder females and pullet-breeder males. So far, in 2004, no British breed clubs have followed this path.

Double mating is not only concerned with plumage pattern: there are some breeds where double mating is applied to comb shape and plumage quality. The rest of this chapter explains the main applications of double mating in use.

DOUBLE MATING FOR COMB SHAPE

Double mating is applied to breeds with large single combs that should be upright on males and flop over on females; this mainly applies to Anconas, Andalusians, Leghorns, Minorcas and Spanish. Females with upright combs are useful for producing the best exhibition males, and floppy-combed males for exhibition females. Most British strains of Leghorns and Minorcas are essentially pullet-breeders, and their cockerels are usually good for one or two shows just as they reach maturity. A month or so later these cockerels are said to have 'gone over'. Environmental temperature plays a part as well, with higher temperatures encouraging comb growth and hence the tendency to flop over.

DOUBLE MATING FOR PLUMAGE QUALITY

Double mating for plumage quality and quantity mainly applies to Dutch bantams. Prize-winning Dutch males should have a wealth of

ABOVE: *White Dutch bantam male. An excellent show bird with profuse 'furnishings' – neck, saddle and tail plumage.*

White Dutch bantam male. A pullet-breeder male with less profuse plumage, especially on the saddle, to produce shapely show females.

saddle hackles, large fanned tails and full neck hackles. Winning females should have rather tighter plumage, especially on their backs and around their thighs and bellies. They still have large fanned tails, the combination giving a wasp-waisted appearance. As Dutch are a relatively new breed at British shows, the British Dutch Bantam Club having been formed in 1982, there are even several leading Dutch breeders here who do not realize that double mating is applicable for consistent success. Many breeders are still at the stage of winning with one sex only, and do not understand why.

The females needed to produce the best show males are those with thicker body plumage on their backs. They completely lack the elegant style of show females, and look rather too large. Pullet-breeder males have much reduced saddle hackles and as a result often look as if their legs are too long; they may actually have the same length legs as show males, but they seem longer because they are more visible.

Dutch bantams are available in a wide range of colour varieties, some of which need to be double mated for plumage colour and/or pattern. These aspects are considered for each colour variety below.

Black-Plumaged Birds with Yellow Feet

This plumage colour mainly applies to Black Leghorns/Italieners, Black Plymouth Rocks and Black Wyandottes. It also could be applied to Black Dresdners (a German breed; there are none in the UK at time of writing), Black Malays and other Asian hard-feather breeds (though the colour is not critical with these, and it may not be worth doing), Black Cochins, Black Pekins (the shanks and feet are largely covered by feathering, so can hardly be seen) and Black Japanese (Chabo) (they have very short shanks, in fact hardly visible).

As both sexes are the same as far as the desired colour combination is concerned, it is surprising to many people that double mating is necessary. The reason is that for any given genetic make-up, the females are darker than the males. The colour spectrum involved here ranges from solid black plumage (as required), but with more or less black shading on the scales of the shanks and feet (not required) to – at the light end – perfect bright yellow shanks and feet, but with less-than-solid black plumage. This might range from white underfluff to white in wing and tail feathers, and sometimes even white spots on the general body plumage. Only the most well organized breeders should attempt to keep both cockerel- and pullet-breeding strains.

Cockerel Breeding: The best males available, with bright yellow feet and sound black plumage in all parts (also correct in type, size, comb and so on for whichever breed) should be mated with females that are 'extra black', usually extending down to a lot of black in shanks and feet. There should still be some yellow showing, even if only the soles of the feet. Preferably avoid black pigment in their combs, faces and eyes.

Pullet Breeding: The best females available, also with bright yellow feet and sound black plumage in all parts, should be mated with males that have white undercolour. White spots on the body plumage is a step too far, as are very light or greenish eyes.

Black-Plumaged Birds with Large White Ear Lobes

This colouring and appearance mainly applies to Black Hamburghs, Black Minorcas and Black Rosecombs; it partly applies to Black Spanish (these have completely white faces, so the problem of white spreading into the faces of the other breeds is not a problem with Spanish). There are other black breeds with smaller white lobes, such as Black Dutch bantams, Black Rheinländers (a German breed) and Black Minorcas as bred in the USA, which have much smaller lobes than British Minorcas.

As with the black birds with yellow feet described above, this is another case that does not seem to need double mating, as both sexes are the same. Indeed, many breeders do stick

Close-up of a Black Rosecomb bantam male.

Close-up of a Black Minorca bantam female.

These two breeds are required to have larger ear lobes in the UK than in other countries. Many breeders of these varieties are competitive exhibitors and regard these difficulties as 'part of the fun'. Judges respect the skills involved, so reward good birds with top awards.

to a single strain, especially with Minorcas, where the large single-comb situation as described above also applies, and nearly everyone concentrates on pullets anyway.

The problem with white ear lobes is that the largest-lobed pullets, the ones that will win at the shows, are bred from males that quickly develop white in their faces and eyelids. Apart from the Spanish of course, this is a show disqualification. More reliable show males, those that will keep tidy lobes for a whole winter show season, will be bred from smaller-lobed hens. Whatever breeding plan is used, the best – and for 'best' read 'largest lobed' – Minorcas and Rosecombs seldom last for many shows as after a few months the lobes will fold up or droop. Large-lobed hens seldom go white in face and lobes, or if they do, it will not be for a year or two. Both breeds are well known for their short show life, which may account for them not being as popular as they once were. From

1900 to 1939 Minorcas were bred in large numbers as layers, and the short show life did not matter then as the breeders were hatching hundreds every year anyway.

Turning to the plumage, the detail under consideration is the green gloss. The females with the greenest-black plumage can be bred from males with some red or yellow in neck hackles and shoulders, which are obviously useless for showing. To avoid producing any when aiming for show cockerels, the smallish-lobed hens should ideally have more matt-black plumage.

Double Mating for Barred Plumage
This mainly applies to Barred Plymouth Rocks (plus Amrocks in Germany), Barred Wyandottes and Scots Greys. Barred plumage is caused by a sex-linked gene that causes plumage production to stop and start as the feathers grow. Male birds have two X

Cuckoo Marans bantam male. Note the generally lighter top colour often seen on cuckoo and barred males. A more even shade overall would be preferred, but this male could be useful for producing clearly marked females. Some Marans females are too dusky overall.

BELOW: *Barred Plymouth Rock bantam male. Excellent British-type exhibition bird with fine barring and of an even shade over all parts.*

chromosomes on which sex-linked genes are carried, and females only one. The barring gene is dominant, but in this case there is a difference between males with one and two barring genes, in that those with two barring genes are generally very much lighter in overall colour than those with one barred gene and one non-barred gene. There are other, largely unknown, genes involved in the overall shade of barred fowls; therefore most exhibition Barred Wyandotte cocks (for example) will have two barring genes, despite being soundly marked with positive black barring. Hens can have only one barring gene, so it is impossible ever to have very light-coloured barred hens.

In the breeds mentioned above, the ideal is for both sexes to have narrow, sharply barred plumage throughout. The overall shade, the general impression when the flock is viewed from a distance, is for both sexes to be the same, though as is explained above, the natural situation is for the males to be lighter.

110

Thus double mating is needed to reliably produce exhibition birds of each sex, although by no means all breeders entirely follow this method. In the case of Cuckoo Marans ('cuckoo' being a coarse, wide, fuzzy version of barring), standard males are much lighter than females, so double mating is not required.

The same barring gene is responsible for all types of barring from the very fine and rather dark barring on British strains of Plymouth Rock bantams, via the slightly lighter and less narrow barring on large Plymouth Rocks and bantams in Germany and the USA, slightly wider bars still on Barred Wyandottes, then Scots Greys, through to Cuckoo Marans at the fuzzy end of this spectrum. These differences are mainly caused by the rate of feather growth. As said above, this effect is caused by stopping and starting of pigment production as the feathers grow: thus Cuckoos are rapid feathering, with only a few on/off sequences in the time it takes for a feather to grow, whereas Barred Rock feathers take a long time to grow, giving a lot of narrow bars.

Cockerel Breeding

The best available exhibition males will be used as breeding stock. Their body shape, legs and feet, head points and so on should all be as good as possible. All plumage should be as clearly and narrowly barred as possible. Ideally all feathers should end in a black tip. White-tipped feathers are most noticeable on the breasts of adult cock birds. Some cockerels have a few of these white tips, but not enough to prevent them succeeding in their first show season, but they will get much worse the next year. Breeders should pay particular attention to the barring on the wing feathers, and the body feathers should be barred down to the skin, including the fluff. Exhibition males should have neck-hackle feathers of the same overall shade as the saddle hackles.

Ideal cock-breeder females should also be clearly barred throughout, but should be rather darker than exhibition females. This overall darker effect is the result of the black bars being wider than the white bars; thus when cock-breeder hens are seen from a distance they look darker. These hens may have a few completely black feathers, which in this case is not a problem as these hens will not be shown. All barred and cuckoo breeds have either white or yellow legs and feet, in some cases with some black spots or shading. The ideal on Barred Plymouth Rocks and Barred Wyandottes is bright yellow. As with the black plumage with yellow feet breeds described above, cock-breeder barred females are also inclined to have dark-shaded feet. This is related to the depth of colour required on exhibition barred males. It does not give dark feet on males, or at least not the darkest of the males, which still have two barring genes. It may occur on males that have one barring gene and one non-barring gene, but such males are usually much too dark, with solid black feathers all over the place, and will be rejected anyway.

Pullet Breeding

Those cockerels that look too light from a distance to be showable will be the ones needed to produce the best exhibition females. Do not go too far on this, however, as the barring should still be clear and distinct throughout. The black bars are narrower on pullet-breeder males, but they should still be present; the weakest barring will probably be on the neck hackles. As with all barred birds, the wing feathers must also be clearly marked, although lighter than exhibition males.

Obviously the best available exhibition-quality females will be used to produce the next generation of champions. Some generally good pullets may have a little dark shading on their scales; they might win some awards in their first season despite this fault, but in later life they will do even better, as scales are moulted as well as feathers.

All barred birds will fade in the sun, especially on their backs. Thus birds with crisp black bars just after the moult may be found to have dull brownish-black bars after a few months. Shaded runs will help to minimize this fading.

DOUBLE MATING FOR THE BLACK-RED/PARTRIDGE PATTERN

This pattern mainly applies to Araucanas, Gold Partridge Dutch bantams, Brown Leghorns, Modern Game, Old English Game and Yokohamas.

The fine details of colour and shade in this plumage pattern differ a little from one breed to another, and novice breeders should consult experts for fine details of each breed. For example, the neck hackle feathers on Black-Red Modern Game males are solid gold, whereas Brown Leghorn male necks have a distinct black centre stripe on each neck feather. Other aspects of double mating, such as the combs of Leghorns and the plumage quality and quantity of Dutch bantams, must be considered as well, as explained in the relevant sections.

Cockerel Breeding

Although the exact shade of the neck and back feathers varies from one breed to another, all exhibition males of this pattern are required to have solid glossy black breasts, underparts, legs and tails. It is also important that they have a well-defined 'wing bay'. When the wings are folded normally, the parts of the secondary flights showing are a bright chestnut colour. They are sometimes rather dark and peppered with black.

The females needed to produce these exhibition males often have lighter coloured backs than exhibition females. Cock-breeder partridge hens are also often 'rusty-' or 'foxy-winged': this is a patch of solid chestnut feathering, and is the female genetic equivalent of the solid wing bays needed on the cocks. As partridge hens have chestnut breasts, in these foxy hens it looks as if the breast colour has spread up on to their wings. The wings of exhibition females should be as their back – brown with fine black peppering. The breasts of cock-breeder females are sometimes a shade lighter, or less red, than the rich salmon-pink shade of exhibition hens, to ensure solid black breasts on show cocks.

Pullet Breeding

For each detail described above for cockerel breeding, the opposite is needed to produce exhibition pullets. All of the back and wings of winning females are brown (the exact shade is different for each breed) with fine black peppering – and definitely no foxy wings! To be sure of this, pullet-breeder cockerels are rather weak in the wing bay, and this part of the wings will have black mixed in with the chestnut/bay colour.

As might be expected, to obtain nice salmon breasts on show females, pullet-breeder males have some brown splashes mixed in with the black on their breasts. Welsummer males were standardized with this mixed breast colour to avoid the need for double mating in this breed.

Exhibition females of all these breeds have more or less solid black centre striping on their neck feathers, and to obtain this, pullet-breeder males have heavier neck striping than exhibition males. As the males of some of these breeds have clear necks, with no striping, it is these females, Modern Game bantams for example, that have less defined striping.

DOUBLE MATING FOR SILVER DUCKWING/SILVER GREY/SILVER PARTRIDGE PLUMAGE PATTERN

This patterning mainly applies to Silver Duckwing Araucanas, S/D Leghorns, S/D Modern Game, S/D OEG, S/D Yokohamas, Silver Grey Dorkings and Silver Partridge Dutch. Double mating does not apply to Duckwing Shamo, Ko-Shamo or other Oriental Game breeds, because these breeds are 'wheaten bred', with 'silver wheaten hens'.

This is the silver version of the black-red/partridge pattern, so when double mating is used the same considerations apply. The term 'duckwing' has been used for centuries by British Game Fowl breeders, and is derived from the band of metallic blue feathers on the cocks of this variety, allegedly similar to that of wild Mallard drakes. The Dutch Bantam Club has decided to follow the more logical term 'silver partridge', as used by their colleagues in The Netherlands. Silver Grey Dorkings were patterned more like Birchens (*see* below) before the shows started in 1845; the female plumage colour soon changed, but the name remained.

Cockerel Breeding
Mate exhibition males with lightish-breasted, foxy-winged females.

Pullet Breeding
Mate exhibition females to cockerels with black-and-white mottled breasts.

DOUBLE MATING FOR BIRCHEN AND BROWN-RED PLUMAGE PATTERN

This patterning mainly applies to Modern Game (and Old English Game in the USA). Note that Brown-Red and Grey (= Birchen) Old English Game are not double mated in the UK, because the breeders and judges are

Pile Old English Game female showing some colour on back and wings. This is necessary to produce rich-coloured males.

concentrating on body conformation, and do not care about the details of plumage colour or pattern. American breeders are much more interested in correct colour and markings.

These are the silver and gold versions of the same plumage pattern. There isn't much difference between males and females of these varieties, so this is another case where the need for double mating is not obvious. They are mostly black, with silvery-white (Birchens) or lemon-gold (Brown-Reds) neck and saddle hackles, and lacing extending from the throat to about halfway down the breast. It is the amount of breast lacing that causes the problems.

Cockerel Breeding
Mate exhibition males with under-laced females, possibly with no breast lacing at all.

Pullet Breeding
Mate exhibition females with over-laced males. In many cases the lacing extends all the way down their legs.

DOUBLE MATING FOR LEMON BLUE AND SILVER BLUE

This colour and pattern mainly applies to Modern Game and Old English Game, as Birchens and Brown-Reds above. This is the same pattern as Birchens and Brown-Reds, but with blue replacing the black. The same considerations apply.

DOUBLE MATING FOR THE PILE PATTERN

The pile pattern mainly applies to Pile Araucanas, Pile Dutch bantams, Pile Leghorns, Pile Modern Game and Pile Old English Game. The name 'pile', spelt 'Pyle' in the USA, is derived from the heraldic term for a triangle or wedge shape that has one side along the top, narrowing down to a point near the bottom of a shield. In poultry terms, it is (nearly) a variation of the black-red/partridge pattern where all the yellow, orange, red and brown parts remain the same, but all the black parts are changed to white. I say 'nearly' because while this is true of pile males, it is not completely true of pile females.

Standard pile females are required to have clear white plumage except for salmon-pink breasts, and yellow with white centre-striped neck feathers. Remembering that partridge females have brown backs and wings with fine black peppering, if it was simply a case of black being changed to white, the backs and wings of pile females would be some brown or reddish shade with fine white peppering. This is why double mating is necessary to produce correctly patterned birds of both sexes.

Cockerel Breeding
As with all cases where double mating is employed, the males used for breeding are the best available exhibition birds. It is widely accepted that the most perfect examples are seen in Modern Game bantams. The best of them have solid rich orange necks, saddles, wing bows and wing bays contrasting with pure white everywhere else. Expert advice will be needed to discover how such birds are 'created'. Breeders of Pile Araucanas, Dutch, Leghorns and/or OEG should also seek the advice of Modern Game breeders for this colour.

The females needed to produce such richly coloured cockerels are called 'rosy piles', and are similar to the above description of what a pile female would be if they were simply a white replacing black version of partridge females. In some strains 'wheaten pile' females are the norm. These are a light salmon-pinkish shade all over, except for white flight and tail feathers.

Pullet Breeding
Again, the best coloured pile females are seen on Modern Game bantams. They have strongly coloured breasts and clear white backs and wings. The contrast looks too dramatic to be real. Again, consult an expert to learn how it is achieved.

To breed females like this, males are needed with very poor wing bays, possibly with nearly clear white wings. This is the male equivalent

of white backs and wings on females. The neck and saddle hackles have clearly visible, white centre stripes, and the coloured parts of these feathers are a shade darker than exhibition males.

DOUBLE MATING FOR THE PARTRIDGE/PENCILLED PLUMAGE PATTERN

This patterning mainly applies to Dark and Gold Brahmas, Partridge Cochins, Partridge Pekins, Partridge and Silver Pencilled Plymouth Rocks, and Partridge and Silver Pencilled Wyandottes. Also, though not standardized at the time of writing in 2004, there are blue versions of this pattern. Those most seen are large Blue Gold Brahmas and bantam Blue Partridge Wyandottes. If the poultry fancy continues to thrive, probably more colour variants will be developed.

Newcomers to pure breeds of poultry are often confused by the inconsistencies in the naming of some plumage colours and patterns. Most of the varieties were named a century or more ago, and the breed clubs are in no hurry to change. After a few years in the fancy most people learn and accept that Partridge Old English Game hens are a different pattern from Partridge Wyandotte hens. They also learn that Partridge Wyandotte and Silver Pencilled Wyandotte hens are different colour versions of the same pattern; and even that Silver Pencilled Hamburghs are yet another different pattern, despite having the same name.

This pattern is sometimes called 'Asiatic partridge', as it is mainly associated with breeds that originated from China, as were Brahmas, Cochins and Pekins. Plymouth Rocks and Wyandottes are American breeds, but Brahmas and Cochins were used to make them. All of these breeds are profusely feathered, rounded-looking birds. The only tight-feathered light breed with this pattern is the Drentse Hoen, a rare breed from The Netherlands.

Males of this pattern are much the same as the other black-red/partridge pattern or its silver equivalent, silver duckwing. The females are very different. They are the same pattern ideally) all over, with each feather having concentric fine black lines around the feather, over a gold or silver ground colour. The exact shade of ground colour and the precision of the markings varies from breed to breed. Some need double mating more than others. Partridge Wyandottes as bred in America are darker and less finely marked than those in Britain. Double mating is not essential in the USA, but in Britain it is so essential that cock-breeder and pullet-breeder strains have become separate varieties in all but name. As far back as 1926, Wyandotte breeders in Germany decided to face this reality and avoid confusing novices, so they changed the names to make them separate varieties.

British cock-breeder partridge =
    German rebhuhnfarbig
pullet-breeder partridge =
    braungebändert
cock-breeder silver pencilled =
    silberhalsig
pullet-breeder silver pencilled =
    dunkel

Cockerel Breeding

The best available exhibition males are used for breeding; they must obviously be correct for comb, breed type, weight and so on for the breed. Ensure that the breast and underparts are solid, glossy greenish-black. Exhibition Partridge Wyandotte bantams in Britain have lighter coloured neck and saddle plumage (orange-yellow shades) than other partridge breeds, which are darker, nearer to orange-red shades. In the silver version of this pattern, as on Silver Pencilled Wyandottes and Dark Brahmas, the top colour must be pure silvery white to contrast sharply with the black centre striping.

Cock-breeder hens should have clear and sharp neck-hackle centre stripes, but otherwise are very fuzzily marked. From a distance they appear generally darker in colour than equivalent pullet-breeder females. These females are never shown, except for the special classes

Silver Pencilled Wyandotte bantam, an exhibition male with a solid black breast.

provided at the Partridge and Pencilled Wyandotte club shows. Some judges have claimed that it is impossible to judge cock-breeder hens or pullet-breeder cocks because they are not standardized. I cannot see the problem myself, as these breeding birds can easily be assessed for type and head points.

Pullet Breeding

All acceptable exhibition females can be used for breeding. The natural tendency is to breed from every hen you have; however, breeders should be organized and self-controlled enough not to breed from hens with inheritable breed faults – badly shaped combs, for example. Sell these hens as pets/layers to customers who are not interested in breeding; do not keep them as broodies or layers as there is too much risk of their eggs being incubated by mistake.

Pullet-breeder cockerels are very obvious in Partridge and Silver Pencilled Wyandotte bantams as they have spotted breasts: brown

115

Silver Pencilled Wyandotte bantam,
a pullet-breeder male with a spotted breast.

Silver Pencilled Wyandotte bantam, an exhibition
female with fine, clear markings. A cock-breeder
female would appear generally darker overall
with less clear markings.

spots on Partridge, white spots on Silver Pencilleds. The neck hackles of pullet-breeder Partridge Wyandotte cockerels are more orangey than exhibition Partridge Wyandotte males.

DOUBLE MATING FOR PENCILLED HAMBURGH PLUMAGE PATTERN (AUTOSOMAL BARRING)

This patterning mainly applies to Gold Pencilled and Silver Pencilled Hamburghs (Hamburghs are called 'Hollandse Hoenders' in The Netherlands). It could possibly also be applied to Pencilled varieties of Friesians.

The Pencilled Hamburgh pattern is very different from the pattern of Silver Pencilled Wyandottes. Experienced exhibitors know this, but it has confused many novices. The pattern on Hamburghs has been named 'autosomal barring' by scientists. Early researchers

in the genetics of poultry plumage patterns were interested in the differences between Pencilled Hamburgh hens (that have finely barred feathers), and sex-linked barring as seen on Barred Plymouth Rocks.

Full descriptions of this pattern are given in the poultry standards books of all countries. It is very obvious that the great difference between the male and female pattern is likely to be a source of problems when breeding them. Exhibition females are finely barred all over, except for the head and upper neck, which is clear gold or silver. Exhibition Pencilled males, despite their name, have hardly any pencilling at all: their main colour is clear gold or silver with some irregular wing markings, a little pencilling on their thighs, and tail feathers that are almost solid black except for a fine edging of gold or silver.

Cock-breeders and pullet-breeders are very different from each other, and should never be

Silver Pencilled Hamburgh bantam, exhibition male with clear white body and black tail. Each tail feather should have a narrow white edging, which is very seldom seen. Nineteenth-century Pencilled Hamburgh breeders were also sometimes tempted to cheat, like their friends, with Spangleds: good tail feathers would be collected, and when needed would be stuck in the hollow base of a faulty marked feather which had been cut off about 2in (5cm) above the visible root.

...mined. Most fanciers concentrate on pullet-breeders; to the extent that many poultry keepers would not be able to correctly identify a cock-breeder hen if they saw one.

Cockerel Breeding

Exhibition males are mated to cock-breeder hens. These have coarser barring, and less of it than exhibition hens. Their breasts are almost clear gold or silver, with perhaps a few black speckles. The tail feathers of cock-breeder hens have irregular markings of a kind likely to be the female genetic equivalent of the desired cock feathers with gold or silver edging.

Pullet Breeding

Pullet-breeder cocks are usually hen-feathered, and have exactly the same plumage pattern as the exhibition hens they are mated to.

Silver Pencilled Hamburgh bantam, exhibition female. A cock-breeder female would have a virtually clear white breast, and the tail markings would be along the length of the feathers, not bars across as seen here.

117

Silver Pencilled Hamburgh bantam, pullet-breeder male. Note that this is henny-feathered.

DOUBLE MATING FOR THE LACED WYANDOTTE PLUMAGE PATTERN

Note that this only applies to Blue Laced, Buff Laced, Gold Laced and Silver Laced Wyandottes, and does not apply to them in Germany, where what would be regarded as pullet-breeder cocks in Britain, America and elsewhere are exhibited. Double mating does not apply to Laced Polands or Sebrights.

With many breeds of poultry the males are more 'showy' than females, but with Laced Wyandottes the females are the main attraction. They are precisely laced all over, apart from striped neck feathers and the fluff around their bottoms. Exhibition Laced males are still very attractive, but the lacing is mainly confined to their breasts and legs. Their neck and saddle hackles are striped, attractive enough, but not very exciting. The top ('shoulder') part of their wings and the upper back should be clear gold or silver. Double mating is necessary because 'clear-topped' males cannot be bred from the same strain as females with perfect lacing on their backs – or not reliably, anyway. Occasionally a clear-top, coloured cockerel will be bred in a pullet-breeder strain, or a fairly good pullet in a cock-breeder strain. Such birds can, of course, be shown, but caution and accurate record keeping are advised if they are to be bred from.

Cockerel Breeding

The most clearly distinct and separate cock-breeder Laced Wyandottes in the UK are a strain of Gold Laced Wyandotte bantams. At the time of writing, 2004, this strain was only kept in existence by a few breeders in Derbyshire and neighbouring counties. This strain is several shades lighter in ground colour than the pullet-breeders kept by the majority of Gold Laced Wyandotte breeders; in this respect cock-breeders, being really gold, are much prettier than the pullet breeders, which are a dark mahogany shade. This strain is not very good in body shape however, being too tall and slim to be a perfect miniature Wyandotte.

Although colour versions of Laced Wyandottes, both large and bantam, are regularly offered for sale, those wishing to establish a proper cock-breeder strain should make enquiries among recognized experts to ensure they get genetically reliable stock. There are some excellent Buff Laced bantams with rich orange hackles and backs, and a few good strains of cock-breeder Silver Laced in both large and bantam. Blue Laced in both sizes, large Buff Laced and large Gold Laced do not exist at the time of writing as truly separate cock-breeder and pullet-breeder types.

Cock-breeder hens should have incomplete lacing on their backs, and may have a lot of 'mossy' peppering in the centre of their back feathers. This peppering is probably the female genetic equivalent of the male hackle-feather striping.

Pullet Breeding

In all colours of Laced Wyandottes, exhibition females should be mated with males that have heavier hackle striping than exhibition males, and also have (sometimes indistinct) lacing over their shoulders and back.

As discussed in the section on the genetics of blue plumage, the breeding of Blue Laced Wyandottes will produce Gold Laced and Buff Laced Wyandottes, of sorts, as well. These Gold and Buff Laced birds can be of use in breeding more Blue Laced, may win some prizes at the smaller shows, and will certainly be attractive enough to find a customer. However, they will be carrying a number of different genes for colour that could introduce undesirable variability into an established strain of Gold or Buff Laced.

Pullet-breeder Gold Laced Wyandottes have been bred with a dark ground colour, more mahogany than gold, because birds of a lighter shade often have very light feather shafts, which detracts from the clear laced pattern required.

Blue Laced Wyandottes are also bred with deep mahogany feather centres. The British Laced Wyandotte Club requires the neck and saddle feathers of Blue Laced males to be as near as possible the same shade as the centres of the breast feathers. Those with brighter, lighter breast-feather centres almost always have bright orange or even yellow hackles. In all cases the centre stripes of Blue Laced Wyandotte hackle-feathers should be blue. The lighter-coloured, orangey-hackled Blue Laced Wyandotte males are considered correct in Germany and The Netherlands, and since most, if not all, of the stock of these colours in the UK originally came from Germany and The Netherlands, it is no wonder that the lighter shades continue to appear.

Many fanciers, especially the British, seem to imagine that keeping 'just' Laced Wyandottes counts as being a 'specialist breeder'. There are four Laced colour varieties, each in large fowl and bantam, which brings it up to eight varieties. If truly distinct cock-breeder and pullet-breeder strains were established for all of them, this would make a total of sixteen strains of Laced Wyandottes. If enough birds were to be kept to maintain the quality of each strain, this would add up to a poultry-keeping establishment much larger than is realistically possible for any hobbyist. There should be more than enough interest for anyone in establishing just one or two strains that are a significant advance on the stock kept by other breeders.

A lot of Laced Wyandottes, especially large fowl, are sadly lacking in true Wyandotte body shape and plumage formation. This is a result of generations of breeders concentrating on lacing for over a century. Novice breeders of Laced Wyandottes, or a lot of other breeds, may think there is not much they can do to improve on existing champions. Nevertheless, although the general quality and numbers of pure breed poultry has improved a lot since they were all facing extinction in the nineteen-sixties, there is a lot more progress to be made by anyone with a focused approach.

CHAPTER THIRTEEN

Selection and Breeding Systems

All breeds of domestic poultry have been artificially created by selective breeding. The general scientific opinion is that all our many diverse breeds were developed from the four species of jungle fowl, mostly the Red Jungle Fowl, found wild in South-East Asia. There has been some debate about additional, unknown species supplying the blue eggshells of South American Araucanas and the very ancient and very large Malayoid and Cochin-type breeds, but this does not change the basic situation. Domestic livestock breeds are all more or less artificial, and without constant selection will tend to revert to an ancestral type.

The need for constant selection is very obvious with some of our popular exhibition types. One does not need to be a genetics expert to realize that strains of perfectly laced Sebrights, heavily feathered Pekins, or tall, refined Modern Game would degenerate after a few years of random breeding. In this case this is assuming that there has not been any crossbreeding, just a failure to select the very best. Sadly this situation does not require much imagination, it happens all the time. People often buy a trio of a breed, and in subsequent years do not breed enough of them to have any choice. Almost every pullet hatched is kept for breeding, no matter how good, bad or indifferent they are. If they still have the breed a decade later, the flock will still be recognizable as whichever breed it is, but minor faults will have become very serious, certainly bad enough for them to be of no interest to any halfway serious potential buyers.

This type of degeneration is most obvious on highly developed exhibition breeds, but it can affect utility breeds as well. A strain of utility Light Sussex might be maintained with perfectly appropriate attention being paid to egg production, growth rate and table conformation, but appearance may be ignored by utilitarian breeders. Breed faults such as white ear lobes and incorrect plumage pattern could get so bad that such a flock could lose credibility as being worthy of the name 'Sussex' at all. It is therefore important for all livestock breeders to inspect carefully every generation of stock bred, and to reject any with significant faults. This is necessary just to avoid a strain getting worse; improvements need even more intense selection.

There are several regular breeding systems used, including inbreeding, line-breeding, outbreeding and upgrading; all can be useful in the right circumstances. Whatever type of breeding programme is to be employed, more progress will be made if a lot of chicks are hatched to give plenty of choice. More chicks of patterned varieties need to be hatched than of self blacks or whites of the same breed. Established inbred strains need not be bred in such great numbers as a variable, recently outcrossed stock.

Some kind of scoring system is best employed if a large number of young birds have to be assessed for several characteristics. There may be a few birds that are generally excellent, and some complete rejects. Most will be good in some respects, poorer in others. Breeders

Successful exhibitors rear large flocks, in this case of Indian Runner ducks, from which only the best will be retained for future breeding and showing.

will need to stick to minimum acceptable quality standards for each aspect; for example, a rose comb with an up-turned leader is out, no matter how good the bird is otherwise.

INBREEDING AND OUTBREEDING IN GENERAL

There are several different systems of both inbreeding and outbreeding employed in breeding systems, but in one key respect they can be said to give one of two general results. Thus, inbreeding produces more homozygous stock: a batch of chicks from an inbred strain will usually all look the same because both parents will be carrying many of the same genes. Outbreeding, on the other hand, produces more heterozygous stock. Because one parent is from one strain and the other parent is from another strain, or even another breed, the stock bred will have different genes at many chromosome loci. In the very special case of commercial hybrids this can still result in a large number of identical chicks being produced, because all the fathers of the chicks were from one very inbred strain and the mothers were from a different very inbred strain. In other circumstances, more normal for amateur breeders, strain or breed crosses

are used to introduce more genetic variability into the flock.

In the course of developing a breed – an additional colour variety of a breed, or an expert exhibitor's particular prizewinning strain – outcrossing is the first part of the process. A logical breeder will have worked out in advance the final objective, and will have chosen the initial stock required. This can be regarded as being similar to a chef devising a recipe and choosing the ingredients. If more than two varieties or strains are involved, then this initial crossing process will take more than one year. Selection and subsequent inbreeding to fix the selected characteristics can only start after all the necessary genes are present in the breeding flock. Breeders have a saying for this: 'You can only breed from the birds you have.' This might be updated to: 'You can only re-combine the genes you have.'

Practical breeders, especially amateur fanciers, may not know exactly all the genes they are concerned with. Of course, most will know about some of them, especially those for plumage colours and patterns that are well documented in the reference books. For other characteristics, the exact size and shape of tail feathers for example, breeders will know the tail formation they require on their breed, but will not know which genes are involved; they just select for what they see.

Whether all the genes are known or not, livestock breeding is all about re-combining genes. Most fanciers have thought something along the lines of: 'If only I had the comb of that one, the tail of another one and the markings of the third, all on one bird.' This is what they are trying to do in their breeding. They should realize that if a particular desired characteristic is not present in any birds in the flock, it is not likely to appear by magic. Yes, mutations do happen, but the right mutation is not likely to happen. Large Laced Wyandottes, especially Buff Laced and Blue Laced males, have the wrong type of tail – long and narrow instead of short and fanned – to give the broad saddled appearance seen in the other varieties. Some progress would no doubt be possible by more careful selection within the existing stock, but major improvement will only be achieved by whoever is brave enough to try a cross (with a White Wyandotte?), and then undertakes the long process of recovering the lacing.

The process called 'adding a gene' is used by geneticists for the process commonly used to create a new colour variety of an existing breed. As an example, suppose a breeder of Welsummers thought it would be a good idea to make a blue-red version of the breed. Obviously the main stock to be used would be his/her existing strain of normal Black-Red Welsummers, as the new variety to be made would be Welsummer in every respect except for the addition of the blue gene from somewhere. The first test of the breeder's skill is his/her knowledge of the options available, and making the right choices. Blue-red Leghorns might seem a good idea at first glance, but their white ear lobes and white eggshells will introduce long-term problems. Blue-Red OEG might be safer, if a yellow-legged strain is available, but this does not help the brown eggs much. Blue Laced Barnevelders, only available in Germany at the time of writing, may seem a less obvious choice because of the different plumage pattern, but will probably be easier in the long run.

The second part of adding a gene, and most other breeding programmes, is an extended period of selective breeding to eliminate almost every trace of whichever breed was used for the cross except for the blue gene. In each subsequent generation the best Blue-red birds would be mated with a normal Welsummer. In this example, it is very important that the new Blue-Red Welsummers lay dark brown eggs as well as look correct.

Completely new breeds are not very welcome in the British poultry fancy, as we have more than enough to keep going already. Imported breeds are a different matter, because they are already well established in their country of origin. Should anyone decide to make a new breed, however, the process is essentially the same, but more complex in the early stages as several different breeds will presumably be needed for all the desired features.

The rest of this chapter is devoted to explaining some of the terms used in livestock breeding.

CROSSBREEDING

This is the crossing of two different breeds or varieties. In conventional poultry breeding terms, this term was mainly used in the 1920 to 1960 period when 'first crosses' were widely used commercially before hybrids were introduced. For example, Indian Game cocks were commonly crossed with medium-heavy-breed hens (Rhode Island Reds, Light Sussex, Plymouth Rocks) to produce good quality table birds when this was part of general farm production. In this type of crossbreeding none of the crossbred birds would be used for further breeding. All crossbred cockerels would be reared for the table, pullets as well with Indian Game crosses. Other crossbred pullets would become layers, but not breeders.

SEX-LINKED CROSSBREEDING

This procedure is the same as the above, but making use of sex-linked genes to give clearly different male and female day-old chicks. The most popular traditional cross on British farms was Rhode Island Red males with Light Sussex females; this gave pale yellow cockerel chicks and orange-shaded pullet chicks; the cockerels grew up to be roughly the same colour as pure Light Sussex cockerels except for some orange feathering on their shoulders. Their growth rate and conformation was a long way short of modern broiler chickens, but was considered good by the standards of those days. Pullets from this cross looked roughly like New Hampshire Red females. They were good layers, but perhaps bigger than ideal for really economic production.

This is called the 'gold × silver mating': it only works with gold male and silver female matings. If done the other way round, all the chicks produced will be silvers. It does not affect the black parts of plumage patterns, these being determined by the other genes relevant for the pattern. Amateur breeders who are more interested in home egg production than exhibition sometimes try sex-linked crosses using Welsummer cocks as gold males because they will be carrying genes for dark brown eggs. If these are crossed with suitable silver females, the resulting chicks will have some partridge markings on their down over their silver or gold ground colour.

For these purposes 'silver' does not mean 'white'. Pure white plumage is caused by different genes, and does not work in sex-linked crosses. Silvers do include the following plumage varieties:

Birchen/Grey, as Norfolk Grey, Old English Game, Silver Sussex
Light/Columbian, as Brahma, Sussex, Wyandotte
Silver Duckwing, as Leghorn, Old English Game
Salmon, as Faverolles
Silver Laced, as Wyandotte
Silver Pencilled, as Plymouth Rock, Wyandotte
Silver Spangled, as Hamburgh

The other plumage colour pair of genes commonly used for sex-linked crossing is barred and non-barred. Barring in this context includes both fine barring as seen on Plymouth Rocks, through to fuzzy 'cuckoo barring' as seen on Marans. It does not include 'autosomal barring' as seen on Brakels, Campines and Pencilled Hamburghs.

Non-barred males must be crossed with barred females for this mating to work, and it can be used in conjunction with the gold × silver-type mating. Several utility-minded hobbyist breeders have bred Welsummer × Marans crosses over the years to produce dark brown egg layers. Day-old cockerel chicks are mostly black, except for yellow spots on their heads. Most breeders will kill all these cockerel chicks as they hatch, but if any are reared as meat birds they will be dark cuckoo barred, similar to Marans hens. The day-old pullet chicks from this mating are mostly black, some showing a brownish tinge; when mature, they look very much like the commercial 'Black Rock' commercial hybrids, mostly black plumage with brown head and neck.

AUTOSEXING

The autosexing breeds were developed to include the benefits of having different coloured male and female chicks in a single pure breed. This is therefore not a type of outbreeding or crossbreeding, but as the name 'autosexing' sounds similar, I thought this would be a good place to mention them. Purebred Cuckoo Marans are almost an autosexing breed in that cockerel chicks are usually lighter coloured than pullet chicks, though this is not 100 per cent reliable. The autosexing breeds were first developed at Cambridge University by a team led by Professor Punnett and Michael Pease. They discovered, initially by accident, that much more reliable day-old chick sex distinction was obtained when the barring gene was introduced on plumage patterns other than solid black. Remember, if they did not carry the barring gene, Barred Plymouth Rocks and Cuckoo Marans would both be black.

During the 1920s the team was studying the difference between normal barring (Barred

Silver Wybar bantam male, one of the autosexing breeds. Bantam versions of the autosexing breeds are a very recent development, mostly kept by those with an interest in genetics.

Plymouth Rocks) and autosomal barring (Gold and Silver Campines). Their first variety, the Gold Cambar, was launched in 1930, with Silver Cambars following in 1933. Neither was good enough a producer to attract much interest, so the principle was tried on more commercially useful breeds. There would eventually be a whole range of autosexing breeds, the most successful at the time being Gold and Silver Legbars, bred from Barred Plymouth Rocks and Brown Leghorns with high producing strains of commercial White Leghorns playing a part in Silver Legbars. As things turned out, the autosexing breeds never became as important as had been hoped; they were overtaken by hybrids, and the introduction of efficient vent-sexing of day-old chicks. No one could have predicted these developments during World War II, when autosexing breeds were thought important enough to have their research funded right through the war. Some of the autosexing breeds still exist.

OUTBREEDING OR OUTCROSSING

These two terms mean the same thing, and usually apply when a closed strain is beginning to suffer from the effects of inbreeding, and unrelated stock of the same variety has to be introduced to rectify matters. Those with successful exhibition strains, particularly varieties with precise plumage patterns, do not usually introduce outside stock unless they have to. But sometimes fertility, egg production, hatchability and general vitality get so bad that the strain is in danger of dying out completely unless something is done.

Sensible breeders will introduce the new strain cautiously by crossing just a few of the original stock with the new stock. All hatching eggs will be marked and hatched separately from other eggs, and the chicks clearly identified with wing tabs or leg rings. Only when the outcross birds are fully grown and can be properly assessed for breed points will the breeder decide if the new strain will be introduced to more, or all, breeding pens.

Many breeders maintain close and friendly contact with other breeders of the same variety, and regularly swap birds; this is normal practice with very rare varieties. If such swaps are frequent it results in the individual fanciers' flocks being effectively merged into one large flock, in genetic terms. In the short term this arrangement has the advantage that there is no need to carefully record every egg and chick as described above: each breeder will know that nothing really odd is going to appear. On the negative side, even the combined flock of all the breeders may eventually succumb to inbreeding depression. Where do they all go then?

Upgrading

This term has been traditionally used in tropical countries where highly productive European and American livestock has been imported, but has been unable to survive the climate and diseases in the tropics. 'Plan B' has been to maintain a small herd/flock in the best available location, a cooler mountainside perhaps, which is then crossed, and crossed again with the hardy, if less productive local breed. After a few years it becomes clear which is the most useful combination of breeds. These (¾ imports, ¼ local?) are called 'grades'. The term is also applied to nearly pure crosses used to revive a breed in danger of extinction.

Backcross

When adding a gene, or outcrossing/outbreeding, 'backcross' is a term that can be used for the matings in the second and subsequent years when the crossbred birds are 'backcrossed' to one of the original inbred strains.

THE CLOSED FLOCK

The ideal for a stable poultry fancy and for the continued conservation of all the old breeds is for as many breeders to keep their strains going as long as possible. As said above, if a strain starts to suffer from the effects of prolonged inbreeding, careful outcrossing can be used. As also said above, if there is only a handful of breeders with a particular variety, mixing the few surviving flocks only delays a possible crisis. Therefore it is very helpful, possibly even critical, if each of the handful of

leading breeders keeps the largest flock they can.

All animals normally have two parents, four grandparents, eight great-grandparents and so on, up to theoretically huge numbers if you go back many generations. As is seen with some wild species on isolated islands, populations can continue perfectly happily for many centuries with quite limited populations. Thus it is obvious that it does not matter if the same animals appear several times in the pedigree of an individual, as long as the number of ancestors is not too limited in recent generations.

A major aim of maintaining a closed flock is to maintain a level of uniformity. That is, all the birds in the flock should conform closely to the breed standard so they continue to be successful show birds. However, if a breeder (obsessive? lunatic?) had several thousand birds of one variety, this uniformity would almost certainly break down because the birds would be too unrelated. These two factors therefore suggest there is an optimum size for a closed flock: it must be big enough to avoid inbreeding depression, but small enough to maintain the desired uniformity. It is normal when breeding poultry to mate each cock with several hens. Thus the two limits to consider are: what is the

A group of several colour varieties of Japanese (Chabo) bantams. It is normal practice, in some cases at least, to cross colour varieties within a breed. The Brown-reds and Black-Tailed Buffs seen here may effectively be one large family. Having two standard varieties in one strain makes keeping a large closed flock more viable for hobbyists.

minimum number of cockerels I must add to the breeding population each year? and, what is the maximum number of pullets it is useful to keep each year?

The key number that all hobbyists should aim for is a minimum of six good quality cockerels to be bred and retained for breeding each year. It is not critical how many, if any, old cocks are retained; this will depend on their quality and continued health and fertility. The theoretical maximum number of breeding females before excess variability creeps in is in the region of 800, and this is well outside the size of flock any exhibitor is ever going to breed. Birds retained for breeding should be as good as possible. None of them will be perfect, but all breeders should have a clear idea of the minimum acceptable quality. The more chicks bred each year, the more successfully the strain will be preserved for the foreseeable future. Most hobbyist breeders, including myself, think they are doing well if they hatch and rear thirty young birds each season of each variety they keep. If a strain were to be seriously conserved, there should be thirty chicks hatched from each of the six cockerels retained for breeding, making a total of 180 to be hatched annually. Very few exhibition breeders are likely to do this, but they should appreciate that this is the ideal for conserving strains.

Line-Breeding

This is a form of inbreeding where maximum use is made of one outstanding individual. If the outstanding individual is a cockerel, he is mated in his first adult year with his mother and/or sisters. Next year he is mated with his daughters, the third year with his granddaughters, and so on as long as he lives; if the original champion bird were a pullet, she would be mated with brother, son, grandson. There are obviously possible dangers with such close inbreeding, so this system should only be used with really exceptional birds, and the rest of the flock should not be abandoned. Some sort of balance has to be maintained between the natural desire to breed a lot more champions, and the long-term need to minimize inbreeding.

Cyclic or Rotational Breeding System

When not distracted into line-breeding by a champion bird, the ideal way a large-scale breeder would keep a closed flock going with the minimum of inbreeding depression is by having six separate breeding pens, corresponding to the six cockerels to be bred each year. To ensure the system does not break down, it would be safest to keep at least twelve cockerels, because each 'official' cockerel needs a spare. The inbreeding is slowed down by putting the cockerel bred from pen 1 into pen 2 next year for breeding. Pullets bred from pen 1 are kept as the breeders in pen 1 the next year. This pattern is continued around all six pens, with the cockerel from pen 6 being bred with the pullets in pen 1. Smaller-scale breeders might like to try this system with four breeding pens.

WHY INBREEDING DEGENERATION OCCURS

There certainly are some harmful, even lethal, effects caused by dominant genes, but the majority of 'bad' genes are recessive. Understanding of how genes work is increasing rapidly, but the picture is still far from complete. Many of them make chemicals essential for body biochemistry, including digestive enzymes. If one of these genes mutates, the mutant gene will probably make a harmless, but less efficient, even useless alternative chemical. Non-inbred stock might have one mutant gene, but the normal gene carried at the same 'locus' on the other chromosome will be able to produce enough of the relevant body chemical to perform the relevant function.

All the birds in an inbred strain are likely to have the same genes for many characteristics. They will have the same genes on both chromosomes for many unexpected and invisible characters, not just the breed points for which the fancier is consciously selecting. Some of these genes have very dramatic effects, and it is the end of the line for a strain if one of these becomes widespread throughout the flock. Most of them are less dramatic, however – slightly reduced growth rate, sperm viability or nutrient absorption perhaps. There may be many such genes present in a strain, initially one or two in this bird, some different genes in another. But each generation of inbreeding brings a few more birds into the population with just a few harmful recessive genes, and thus the effects of inbreeding get gradually worse.

Do not let this frighten you from inbreeding completely. Some breeders, none of them well known, have a horror of inbreeding. They regularly buy 'fresh stock', and will proudly advertise 'unrelated trios'. This is fine for utility breeds, because the unrelated pullets bred from the original trio will be better layers for being bred from unrelated parents; but this is not so with purely exhibition varieties.

WHEN TO SELECT?

Selection should be a continuous process. Plumage pattern cannot be assessed very early because juvenile plumage is different from adult plumage, but poor-shaped combs can be spotted quite early on, especially on cockerels. Bantam cockerels are seldom worth rearing as table birds, and will probably start fighting before they are mature. No one else will want obviously faulty cockerels, so they will have to be killed, but there are usually more than enough cockerels left when the

flock has been reduced to the good ones to supply the limited demand for stock cocks.

SELLING THE REJECTS

Breeders must be honest with potential buyers. They should price birds realistically according to their quality, and only healthy birds should ever be sold. Taking Laced Wyandottes as an example, only the 'best of the rest' should be sold at breeding stock prices to people intending to breed and show them. Pullets with much poorer lacing can still be sold with a clear conscience as pretty garden pets and layers, but they should go at pet prices.

ABOVE: *A group of Partridge and Silver Pencilled Wyandotte bantam pullets. There were probably more than this to begin with, and not all of them would have been permanent members of the flock. Faulty combs could be identified early on, but details of plumage pattern can only be assessed when birds are nearly mature.*

Another view of this Wyandotte specialist's runs. Many beginners would be tempted to have a different breed in each ark, but they should resist this temptation to dabble.

CHAPTER FOURTEEN

Health and Disease

For this subject to be covered properly, a shelf full of books this size would be needed, at least one book for each disease. A brief paragraph on each disease may cause more confusion than illumination as there are, for example, several respiratory diseases, all of which have similar symptoms. Many useful single books have been published over the years, which give a good insight for poultry keepers. A veterinary surgeon will need to be consulted to treat most disease problems, but some knowledge will help poultry keepers quickly note that there is a problem; the flock owner can then contact a vet and say 'I think my birds might have ...'. Most of this chapter is devoted to those problems that a normal poultry keeper can control: external parasites (lice and mites), internal parasites (worms) and hygiene awareness.

As regards vets, not all deal regularly with poultry, and as a result may not have the relevant drugs in stock. Experienced members of your local poultry club should be able to suggest the best vets for birds, or failing that, you could phone and ask some listed 'poultry farmers' in *Yellow Pages* whom they use.

If a problem arises for which a vet seems necessary, it is normal practice to make an appointment to take a few sample sick birds into the surgery. Home visits are very expensive. Tell the vet how many birds you have in all, and how many of them show symptoms. Special measures would apply if Newcastle Disease (fowl pest) or avian influenza appeared again, probably including a hotline to call. The authorities would not want birds with these infectious diseases taken anywhere.

GENERAL HEALTH AND HYGIENE

The healthier your birds are, the less likely they are to succumb to diseases. One of the most important factors is clean drinking water in clean containers: you should not expect your livestock to drink water you wouldn't drink yourself. Open drinkers must be emptied and re-filled every morning, especially if there is any chance they might have been polluted during the previous day or overnight by wild bird droppings or rats and mice. It is preferable for all food troughs to be emptied and the food securely stored (from rats and so on) overnight. Chicks, whether under heat lamps or with broody hens, should have food at all times, but they should be in pens that are rat proof, mouse proof and sparrow proof anyway. As described in the nutrition chapter (*see* pages 49–54), all stock should have the correct diet for their age, as much as they will eat of it, and any problems that might depress food consumption should be addressed. Vitamin deficiencies can cause health problems long before the 'classic' symptoms fully described in specialist disease and nutrition text books become apparent. Good housing is vital as well, but it is to be hoped that exhibition breeders will have more than adequate hen houses. The conditions necessary to keep birds perfectly clean for showing should also be healthy.

Biosecurity

Commercial poultry-breeding units are like high-security establishments: very few visitors are allowed, and they must wear a complete suit of sterile protective overalls, boots, and so on. Only day-old chicks direct from the hatchery enter the site, and birds at the end of their production period go directly for slaughter. None of these conditions applies to exhibition poultry-keepers, however, for whom the main point of keeping birds is to take them to shows and bring them home again. Many regular exhibitors have individual pens for their main show birds during the winter show season, and these are a useful isolation measure in case they pick up an infection. Furthermore, a good precaution for new stock bought in is an initial period, perhaps three weeks, in an isolated house.

If it is possible to set up disinfectant footbaths, as used commercially, do so. Maintain poison and/or traps to control rats and mice as standard procedure, and place food and water troughs inside hen houses where they are less likely to be polluted by vermin or wild birds.

Exhibitors should be keeping their birds in cleaner and less crowded conditions than normal domestic poultry keepers, and this should compensate for the extra disease contacts at shows that their birds have to cope with.

Crossing Borders: Quarantine Regulations

Regulations concerning the international transport of livestock are different for each country, or in the case of the European Union, group of countries. They are likely to change over time, so anyone intending to import poultry should contact their national poultry-showing organization and relevant government department (PCGB and DEFRA in the UK) for current rules. Fanciers in Australia and New Zealand have not been allowed to import anything for a very long time.

In general terms, potential UK importers will need to have their stock checked and certified by a vet at the country of origin, and again when they arrive, and again after a specified period. They must be housed inside a building above a

regulation distance from other poultry. This building must conform to size and construction rules, mainly requiring washable inside surfaces. It may be possible to arrange the use of an existing quarantine facility rather than build your own. If regular imports are intended, it may be better to construct the building at the property of a non-poultry-keeping friend or relative instead of your own.

EXTERNAL PARASITES

All domestic-scale poultry keepers are, or should be, constantly checking for and controlling lice, northern mite, red mite, scaly leg mite and other infestations. The first three are most active in warm weather, and scaly leg is particularly associated with certain breeds, especially those with feathered legs. Every fancier has a selection of louse powders, mite sprays and scaly leg treatments in their chicken medicine cupboard. Stock and houses need to be checked regularly, with extra treatments for show birds (before and after each show) and broody hens. War should be declared on external parasites every August, the beginning of the moulting season, so that the flock is free of them on their new plumage.

Lice

These are fairly large, and light brown in colour, and although an infestation may look alarming, they are seldom quite as harmful as mites, and are fairly easy to kill. Clumps of louse eggs appear as greyish clumps at the base of feathers, mainly around the vent area of birds. Sprays and powders kill live lice, but do not kill the eggs. Repeated treatments are needed to kill them as they hatch over several weeks.

Red Mite and Northern Mite

Although they are two different species with different behaviour patterns, they look very similar, so perhaps can be treated as a single problem. They are more difficult to kill than lice, louse powder having no effect. Northern mite live on the birds all the time, whereas red mite spend their days in cracks and crevices in poultry houses, mainly on or near perches and nest boxes. A greyish deposit, mite faeces, can be seen when houses are checked in daylight. Red mite can kill broody hens because they can suck their blood continually.

Some very potent sprays are available for disinfecting houses, but these are too strong for direct use on birds, or even for use in houses with birds inside: the fumes would kill your poultry, so shut them out of the house for an hour or so after spraying. Sprays intended for use on pigeons and cage-birds, the only ones sold at many pet shops, are just as effective on poultry.

Depluming Mite and Feather Mite
These are not as seriously health-threatening as northern and red mite, but can ruin the appearance of exhibition birds. Mite sprays should be effective, but better results for key show birds can be achieved by a full bath with insecticidal shampoo.

Scaly Leg Mite
You will never see these mites, but their effects are very unsightly and painful for the birds. The mites burrow under the leg scales, causing them to be raised and deformed with a characteristic greyish encrustation. If allowed to progress untreated the condition can cause bleeding, and birds can be seen pecking at their own shanks, a sure sign of severe discomfort. Even a modest scaly leg encrustation will justify disqualification at a show. Novice poultry-keepers should seek advice from an experienced person, as they may confuse scaly leg with naturally raised and coarsened scales on old birds.

The traditional remedy was to dip the feet in paraffin heating oil; however, the application of surgical spirit with a cotton bud is more common now. Spraying with mite spray can also be effective, though if the condition has caused bleeding, a gentler product will be more humane. A thick coating of paraffin wax – for example Vaseline – may suffocate the mites. Benzyl Benzoate BP, a cream for skin conditions, is now used by several fanciers to treat scaly leg and other skin problems on poultry.

INTERNAL PARASITES

Most of these infest the intestinal tract, and include caecal worms, hair worms, roundworms and tapeworms. The exception is gapeworm, a species that lives in the windpipe and causes gasping symptoms that could easily be confused with other respiratory diseases. All the intestinal species of worm cause weight loss, often despite a noticeable increase in food consumption.

Suitable treatments are mixed with the food; Flubenvet™ is the usual brand available in the UK at the time of writing. These drugs are available (in 2004) from agricultural supplies centres, but future legislation may restrict this availability to veterinary surgeons. A routine treatment of the whole flock twice a year is recommended, plus additional treatments if worms are seen in the droppings.

Although Flubenvet™ is effective against gapeworm, it might be prudent to have suspect birds checked by a vet, in case it is a different disease.

VACCINATIONS

Commercial pullet rearers customarily give their birds a complete vaccination programme, starting with Marek's disease at the hatchery. Most vaccines are only supplied in 1,000 dose bottles, which is a problem for hobbyists, but smaller quantities may be available in the future, and some of the 1,000 dose vaccines are cheap enough to buy, even if only 100 doses will be used. Consult a vet for current availability, costs, and the age at which they should be administered. Apart from Marek's, the other main vaccines are to prevent Newcastle disease (fowl pest), fowl pox and infectious bronchitis ('IB'). If there is a local problem, vaccines are also available for Gumbaro disease (infectious bursal disease), infectious laryngotracheitis, coryza and *Escherichia coli*.

Apart from the problems hobbyists might have in buying vaccines in small quantities, some fanciers believe that the very nature of the hobby – access to free-range, flocks of mixed age groups, and visiting shows – render the commercial approach to disease prevention completely inappropriate. They say that, unless there is a clear and specific threat, it is best to rely on natural disease resistance. Some specific genes affecting disease resistance are being identified and are now better understood, mainly relating to Marek's disease. This may not be of much help to those who are already committed to conserving a specific variety, but new fanciers might like to know that the Mediterranean breeds are more resistant to Marek's disease than the heavy Asiatic breeds and Silkies.

COCCIDIOSIS

This is an intestinal disease caused by several species of protozoans, all of which are species-specific; thus rabbit coccidia species do not affect chickens. The species of Coccidia that affect chickens are as follows *Eimeria acervulina*, *E. brunetti*, *E. hagani*, *E. maxima*, *E. mitis*, *E. mivati*, *E. necatrix*, *E. praecox* and *E. tenella*, the last named being the most serious. Although there is some overlap, generally each species attacks certain ages of chicken and at specific parts of the intestinal tract. *E. tenella* affects birds between two and twelve weeks of age, with a peak at about six weeks. This is about the age at which young chickens are often moved from their brooding pens (with hens or under lamps) to larger accommodation, possibly outside for the first time. Thus your six- to eight-week-old youngsters, having been growing strongly so far in a nice clean pen, encounter dangerous disease organisms at the same time as they are stressed by a house move and possibly a change of diet, from chick crumbs to growers' pellets as well. Infected chicks look obviously sick, standing with ruffled feathers. They pass diarrhoea at first, including vent soiling, and then show the classic Coccidiosis symptom: blood in the droppings. There can be a high death rate only a week after a sudden heavy infection.

Coccidia species have a life cycle that includes a stage when resistant spores, or 'oocysts', are passed out in the droppings of an infected bird, where they can survive for over a year, being most active in warm and damp conditions. Poultry keepers should assume that all housing that has ever had a chicken in it before has at least some Coccidia oocysts present. Most chickens have some natural resistance to low-level infection, helped by preventative drugs ('coccidiostats') in most chick crumbs and growers' meal/pellets. (Note that there are organic rations that do not have these additives, but the author does not recommend them, having seen more than enough birds die of this horrible disease.) Foods including coccidiostats do not always prevent the disease if there is a high concentration of a virulent strain. Some sample sick birds should be taken to a vet as soon as possible for confirmation of diagnosis, and to obtain the stronger drugs that are available to treat them. These drugs are not included in foods as a routine preventative, as the Coccidia would soon become resistant to them, a problem that readers will know is a concern with antibiotics generally.

MAREK'S DISEASE

This disease is a type of cancer and is caused by a herpes virus, and just as herpes viruses can cause several different effects in people, so they do in poultry as well. The 'classic' form is commonly called 'fowl paralysis' and affects one or both wings and legs. Post-mortem examination also shows tumours on the liver and other internal organs.

There is a considerable variability in bird genetic resistance to Marek's disease, as has been noticed by many poultry keepers when there is a high loss in one breed, and hardly any birds affected in another. Once affected, there is no treatment to cure Marek's, but a (very) few naturally resistant birds will limp for a week or so and then recover.

There is a vaccine, but it is only available in 1,000 dose bottles, and it must be used within four hours once opened; it is given to day-old chicks at the hatchery. It is not viable to vaccinate hen-hatched chicks, as they will have contracted the natural virus as soon as they hatch. Most fanciers accept losses, and hope the survivors are more resistant.

Marek's disease is so widespread in the poultry population that it must be assumed that once a chick has been anywhere near other chickens it will have contracted the virus. They will become more or less resistant according to the genes they carry, and if they ever do succumb to the disease, it may not be for many months. Sometimes it can be set off by other stress factors such as starting to lay, their first show, bullying, or another disease.

AVIAN INFLUENZA AND NEWCASTLE DISEASE (FOWL PEST)

These are two of the most dangerous poultry diseases, so serious that if there is a current outbreak it will be fully reported in all the news media. Both can infect people, avian influenza being more likely to do so, and being much more serious. Any suspected outbreaks must be reported to the relevant government department, which will be operating a slaughter policy to eradicate the problem. However, at the time of writing (2004), representatives from the Poultry Club of Great Britain had met the Animal Movements and Exotic Diseases team from DEFRA to start negotiating exemptions from slaughter where possible for important rare pure-breed flocks. This is because in 2003 there was a major outbreak of avian influenza in Belgium, Germany and The Netherlands, and over 28 million birds died or were slaughtered, causing huge financial losses to the poultry industry. Compared to this, the few hobbyists' flocks of exhibition fowls and bantams that were also lost might seem insignificant. However, some of these birds were very important in historic and genetic diversity terms.

Both diseases show respiratory symptoms for a short period before the death of the birds in the most acute outbreaks. Consult a detailed and up-to-date information source for full details, perhaps the (UK) DEFRA or (USA) USDA web sites.

MYCOPLASMA, INFECTIOUS LARYNGOTRACHEITIS AND OTHER RESPIRATORY DISEASES

These, along with gapeworm infestation as mentioned already in the 'Internal Parasites' section, can easily confuse even experienced poultry keepers. The symptoms can be very similar for all of them: coughing, gasping for air, frothy eyes and swollen sinuses. A vet should be contacted as soon as possible for a correct diagnosis and drugs to treat the problem. Viruses are not usually treatable, but antibiotics may be diagnosed to prevent any secondary infection complications. Only a vet dealing with commercial poultry farms in the area will know the current disease problems. Post-mortem examinations may be necessary, but a vet who already knows there are commercial flocks with one of the diseases locally may save this expense. Vets may require blood samples to be taken for testing.

SALMONELLA, *E. COLI* AND OTHER BACTERIAL INFECTIONS

Often only laboratory examination of infected blood or other samples will identify the disease present, and so the antibiotic most likely to cure it. Sometimes laboratories never do identify the bacterium involved, but fortunately they can often report: 'Whatever it is, such-and-such-a drug will cure it.' This is discovered by antibiotic-sensitivity testing.

Routine blood testing is required for all commercial poultry flocks to prevent consumers' contracting infection via eggs or poultry meat. Small flocks for domestic consumption, plus perhaps some sales to friends and neighbours, are generally exempt from testing regulations. The rules defining 'a domestic flock' (such as maximum number of birds) are likely to be different in each country, and change over time. Exhibition poultry keepers may need to check that their flock is below 'commercial' size.

The Poultry Fancy Worldwide

Every country with an exhibition poultry fancy has a similar arrangement of clubs to administer the hobby. The number of clubs clearly varies with the geographic size of the country and the number of breeders. Some small (in poultry-showing terms) countries are, in this respect at least, offshoots of their nearest large neighbour. The normal structure is as described below.

A national club, federation or council: This may also involve similar hobbies such as fancy pigeons, cage-birds, rabbits and other small livestock. It usually runs a national show, sets show rules, breed standards and a test or other system for approving judges.

Breed clubs: These typically are for all the breeders of a single breed in the whole country. Some clubs cater for two or more related breeds, such as the British Belgian Bantam Club, which covers five breeds. The clubs usually hold several specialist shows for their breeds, publish yearbooks, newsletters and breeders' lists. Many of them maintain close links with equivalent clubs around the world. This is particularly true in Europe where at major shows it is quite common to find breeders from all over the continent discussing the finer points of their chosen favourites.

Larger multi-breed clubs: Britain has the Rare Poultry Society to cover those breeds that are in too few hands to have their own club. The American Society for the Preservation of Poultry Antiquities (SPPA) is similar. Several countries have a Waterfowl Club covering all breeds of duck and geese.

Local poultry clubs, fur and feather clubs and similar: These cover all (relevant) breeds in their area and organize the shows. They usually, in the UK at least, own and erect the show cages, hire the hall, book the judges and so in. In addition, some of them hold regular monthly meetings and arrange bulk-buying of chicken food, vaccines or other requirements.

Agricultural shows (= state fairs in the USA): These include a poultry section and are usually affiliated to their national poultry club. In the UK, the poultry sections of the agricultural shows are often organized by the nearest local poultry club committee for a mutually agreeable financial arrangement. This may also be the case in other countries.

This general structure has been normal since at least the 1880s, and is likely to remain so for the foreseeable future.

Below are all the poultry show organizations I could find. Local clubs have not been listed, neither have any names or contact numbers, as club officials change over the years and many names could be different by the time this book is printed. Consult poultry sites on the Internet or appropriate magazines for current club officials. No doubt some clubs will close down and new ones will start up, but this list gives an idea of the strength of our hobby around the world.

GREAT BRITAIN

Poultry Club of Great Britain: this is the governing body.
National Federation of Poultry Clubs: mainly organizes a very large show at Stafford.

There are about 150 local clubs, including those affiliated to agricultural shows.

Breed Clubs

Ancona Club
Araucana Club
Australorp Club
Barnevelder Club
Belgian Bantam Club
   (covers Barbu d'Anvers, Barbu d'Grubbe, Barbu d'Everberg, Barbu d'Uccle, Barbu d'Watermael)
Black Wyandotte Club
Brahma Club
British Call Duck Club
Buff Orpington Club
Cochin Club
Croad Langshan Club
Derbyshire Redcap Club
Dorking Club
Dutch Bantam Club
Faverolles Society
Frizzle Society
Hamburgh Club
Indian Game Club
Indian Runner Club

Japanese Bantam Club
   (Chabo)
Laced Wyandotte Club
Leghorn Club
Lincolnshire Buff Club
Marans Club
Minorca Club
Modern Game Club
New Hampshire Red Club
Old English Game Bantam Club
Old English Game Club, Carlisle
Orpington Club
Partidge & Pencilled Wyandotte Club
Pekin Bantam Club
Plymouth Rock Club
Poland & Poland Bantam Club
Rhode Island Red Club
Rosecomb Bantam Club
Scots Dumpy Club
Scots Grey Club
Scottish Ancona Club
Scottish Game Club
Scottish Pekin Bantam Club
Scottish Plymouth Rock Club
Scottish Rhode Island Red Club
Scottish Rosecomb Club
Scottish Sussex Club
Sebright Club
Silkie Club
Sussex Club
Welsummer Club
White Wyandotte Club
Wyandotte Club

Scots Grey bantam pair. Not classed as a 'rare breed' in the UK, because we have the Scots Grey Club. They are, however, very rare worldwide, and even the UK Scots Grey Club has a limited membership.

Before 1939 several breeds had separate clubs for specific colour varieties, but most of these have amalgamated to a single club, except for Orpingtons and Wyandottes.

Multi-Breed Clubs

Asian Hard-feather Club: covers Asil, Ko-Shamo, Malay, Nankin-Shamo, Shamo, Yamato-Gunkei and related breeds

British Waterfowl Association: also for breeders of wild species

Domestic Waterfowl Club

Goose Club

Midland Old English Game Club: covers Oxford-type OEG

Rare Poultry Society: covers approximately sixty breeds of fowls and bantams

Utility Poultry Breeders Association: to encourage production recording of any pure breed

Turkey Club

Worshipful Company of Poulterers: one of the ancient City of London Guilds

The Entente Européenne

This is a federation of European poultry, pigeon and rabbit organizations. There are over twenty member countries, including Belgium, France, Denmark, Germany, Italy, Luxembourg and The Netherlands. The PCGB is a member, but is less able to take a full part because of British Government quarantine laws.

INTERNATIONAL BREED CLUBS

There have been numerous attempts at forming international breed clubs over the years, mostly with very limited success. Most of the genuine international contacts are in European Union countries, as it is so easy to drive from one country to another, especially between Belgium, Denmark, France, Germany, Luxembourg and The Netherlands. Newsletters are regularly posted around the world, and most breed clubs have a few overseas members. Some of the breed clubs in the USA section of this appendix are called 'World...' or 'International...', but it is

assumed that most of the members are in the USA and Canada. Breeders can now keep in direct contact around the world via the Internet, which reduces the need for formal 'international' breed clubs.

International Poland Club links all Poland and Crested Breeds clubs around the world

BELGIUM

Koninkije Bantam Club van Belgie (KBCB) (Royal Bantam Club of Belgium)

Belgian Breed Clubs

Speciaalclub Voor Het Brakelhoen

Speciaalclub Belgische Hoenders en Dwerghoenders (SBHD)

Also a club for Chabo (Japanese) bantams, name of club unknown

DENMARK

Danmarks Fjerkæavlerforening forRaceavl (DFfR): this is the governing body.

The DFfR is divided into twenty-five districts. There are additional local clubs.

Breed Clubs

In 2002 there were twenty-nine breed clubs in Denmark, including:

Chabos-Klubben

Danske Landhøns Club (for the local Danish Landhen breed)

Landsklubben for Orpington (also covers Australorps and Jersey Giants)

Minorca Club

FRANCE

Société Centrale d'Aviculture de France

Breed Clubs

Poule et Oie d'Alsace Club

Club Français des Races Avicoles Ibériques et Latino-Américaines (Araucanas)

Close-up of a Houdan's head. This French breed is enjoying a revival in its homeland, but remains rare elsewhere. Note the unique 'butterfly' comb, like a cup comb, but pushed forward by the crest. Circa 2000, some excellent Houdan bantams appeared at a few British shows, but they were only kept by a couple of fanciers.

Conservatoire de l'Ardennaise
Association pour la sauvegarde de la poule de Barbezieux
Bourbonnaise Club
Association de éleveurs de races Bourbourg
Brahma-Cochin Club
Groupement National des Éleveurs Bresse-Gauloise Club
Club pour la Sauvegarde de races Avicoles Normandes *
Conservatoire de races Normandes et du Maene *
(* These cover Caumont, Courtes-Pattes, Crève-Coeur, Gournay, Merlerault, Pavilly)
Club Francais de Combattant du Nord

Club des Coucous de Flandres et de Picarde
Amicale des Dorkings de France
Padgalli Estaires Club
Houdan: Faverolles Club de France (also covers Mantes)
Gasconnes Club
Gâtinaise Club
Conservatoire des Races du Nord et de Picardie (Braekels, Hergines)
La Flèche Club Français (also covers Le Mans)
Club de la Limousine
Marans Club de France
Club de la Volaille Meusienne
Bantam Club Français (covers Pictaves)
Club Du Barbu De Watermael

GERMANY

Bund Deutscher Rassegeflügelzüchter
Verband der Hühner-, Groß- und Wasserge flügelzüchtervereine (9,000 members)
Verband Deutscher Rassegeflügel-Preisrichter (judges' association)
Verband der Zwerghuhnzüchter (association of bantam breeders)

In Germany there are (in 2004) seventy-six breed clubs and six regional clubs. There is also a huge number of local poultry clubs, many based at communal 'allotments'.

SV Araucana und Zwerg-Araucana
SV dZ des Barnevelder und Zwerg-Barnevelder-Huhnes
SV d Zuchter des Italiener-Huhnes (Zuchter or Z = Breeder)
SV dZ des Zwerg-Italiener-Huhnes
SV dZ der Zwerg-Langschan (German Langshan bantams)
SV dZ des New Hampshire-Huhnes
SV dZ der Zwerg-Niederrheiner
Orpington Specialzuchter club
SV dZ der Zwerg-Orpington
SV dZ der Zwerg-Reichshuhner
SV dZ des Rhodelanderhuhnes
SV dZ der Zwerg-Wyandotten
SV dZ Seltener Huhnerrassen (Rare Breeds)

*Black German Langshan bantam male. Popular in Germany, but rare elsewhere. Large entries could be seen at some British shows (*circa *1995–2005), but they all came from a small group of keen exhibitors, notably Brian Ward of Bedfordshire.*

HUNGARY

There is a Hungarian Pigeon and Small Animals Federation, but its Hungarian name is not known by the author, neither is the proportion of poultry breeders.

IRELAND

Irish Society of Poultry Fanciers

There are surprisingly few regular exhibitors in the Republic; most of these attend shows in Northern Ireland and join PCGB affiliated breed and local clubs.

ITALY

Federazione Italia Associa Viario (Federation of Italian Avicultural Associations), also local and regional clubs.
Club Italiana della bantam

LUXEMBOURG

Union of Societies in Aviculture for the Grand Duchy of Luxembourg. Because of the small size of Luxembourg and its central location, the fancy is closely allied to that of Belgium, France, Germany and The Netherlands.

THE NETHERLANDS

Nederlandse Bond van Hoender-, Dwerghoender-, Sier, en Watervogelhouders
ANV Dwerghoenderfokkersvereniging
Grote Hoender Club
Nederlandse Hoender Club

Breed Clubs
Nederlandse Ancona Club
Antwerpse Baardkrielen Club (Barbu d'Anvers)
Assendelfter Hoen en Noord-Hollandse Blauwe Club
Barnevelder club
Speciaalclub Belgische Hoenders en Dwerghoenders
Specialclub voor Zeldzame Oorspronkelijke Belgisher Krielhoenderassen (Special club for rare Belgian originated bantams)
Speciaalclub voor fokkers van Brabanters, Kraaikoppen en Uilebaarden (Special club for breeders of Brabanters, Bredas and Owlbeards)
Brahma Club
Speciaalclub voor Brakelhoen
Chabo Liefhebbers Club (Liefhebbers = amateurs or hobbyists)
Cochin club
Drentse Hoen Club
Franse Hoender Club (for French breeds)
Friese Hoenclub
Groninger Meeuwenclub

Hollandse Hoenfokkersclub (Hollandse Hoen = Hamburghs)

Hollandse Krielenfokkersclub (Dutch Bantams)

Nederlandse Java Club (Rosecomb Bantams)

Jersey Giant Club

Nederlandse Kuif- en Baardkuifhoender Club (Polands, kuif = crest)

Lakenvelder Club

Nederlandse Leghorn Club

Nederlandse Minorcavereniging (vereniging = union/society/club)

Nederlandse Plymouth Rock Club

Rhode Island Red Club van Nederlandse

Nederlandse Rijnlander Club

Nederlandse Sebright Club

Nederlandse Sabelpoot Club (Booted Bantams)

Speciaalclub voor Sulmtaler en Alsteiererfokkers

Nederlandse Sussex, Orpington en Dorking Club

Thuringer Baardhoen en Krielen Club

Twents-Hoender Club

Watermaalse Baardkrielen Club

Welsumer Club

Nederlandse Wyandotte Club

Zijdehoen Club (Silkie Club)

Multi-Breed Clubs

Vechthoender Club (Game breeds, Vecht = Fight)

Watervogels Ver. van Gedomesticeerde Watervogelfokkers

There are also numerous local clubs, but no details available.

NORWAY

Norwegian Poultry Club/Norsk Rasefjaerfeforbund Poultry Club

SPAIN

Associación Espanola De Avicultura Artistica

Associación De Criadores De Aves El Francoli (French breeds?)

Associación De Criadores De La Raza Prat

Associación De La Gallina Vasca

SWEDEN

Svenska Lanthonsklubben

Göte Borgs Ortens Rasfjädefäklubb

SWITZERLAND

Bund Schweizerischer Rassegeflügelzüchter

Schweizerischer Geflügelzuchtverband

Schweizerischer Siedenhuhn und Hauben hühner-Züchterclub (Silkies and Crested)

AUSTRALIA

Australian Poultry Exhibitors Council: a committee of representatives from the state associations. Because of the great distances involved, the fancy is mainly organized at state level. There were (in 2002) over 200 local poultry clubs in Australia, forty of which are in the state of Victoria; most of the breed clubs are based in New South Wales, as well as another seventy local clubs.

New South Wales

Exhibition Poultry Association of New South Wales

Bantam Club of New South Wales

Ancona Club of Australia

Araucana Club of Australia

Australorp Club of Australia

Belgian Bantam Club of Australia

Brahma Club of Australia

Leghorn Club of Australia

Old English Game Fanciers Association of Australia

Orpington Club of Australia

Pekin Bantam Club of NSW

Rhode Island Club of Australia

Rosecomb Bantam Club of Australia

Sussex Club of Australia

Queensland

Feather Clubs Association of Queensland

Australorp Club of Queensland

South Australia

Fifteen local clubs

Waterfowl Club of Australia (South Australia
  Branch)

Tasmania

Southern Tasmania Poultry Club

Victoria

Victoria Rare and New Breeds Poultry
  Society

Western Australia

Western Australian Poultry Association plus
  local clubs

CANADA

The hobby in Canada is closely associated with
the USA.
Canadian Bantam Club
There are provincial poultry clubs for Alberta,
Manitoba, Ontario, Quebec, Prince Edward
Island, Saskatchewan and Vancouver.
Canadian Araucana Society
Canadian Plymouth Rock Club

JAPAN

Language difficulties prevent a full listing of
Japanese fancy poultry clubs, but they certainly
exist at national, local and breed club levels.
Almost all the interest is in Japanese breeds:
Chabo, Shamo, long-tails, long-crowers and so
on.

NEW ZEALAND

North Island (NZ) Poultry, Pigeon and Cage-
  Bird Association
South Island (NZ) Poultry, Pigeon and Cage-
  Bird Association

NZ Breed Clubs

Auckland Game Fowl, NI Game, Hamburg,
Orpington, Rock and Australorp,
Wyandotte, Mediterranean, Rhode Island

Red, Silkie, Waterfowl, Pekin, Rosecomb
and Sebright

There are also twenty-four local clubs and specific cage-bird and pigeon clubs in New Zealand.

SOUTH AFRICA AND NAMIBIA

Southern African Show Poultry Organization:
this is the governing body. The affiliated clubs
include Namibia Poultry Club. This sparsely
populated neighbouring country does not have
enough fanciers for more clubs. There are also
nine local clubs in South Africa.

Breed Clubs

Waterfowl Association of South Africa
Old English Game Bantam Club of South
  Africa
Orpington Club of South Africa
Oxford Old English Game Club of South
  Africa

USA

American Poultry Association
American Bantam Association
Society for the Preservation of Poultry
  Antiquities (SPPA)

There is a large number of local and state-sized
poultry clubs.

Breed Clubs

Ameraucana Breeders Club
American Australorp Breeders
American Brahma Club
American Brown Leghorn Club
American Buttercup Club
American Dutch Bantam Society
American Langshan Club
American Silkie Bantam Club
American Sumatra Association
American White Leghorn Club
American Wyandotte Breeders Association
Araucana Club of America
Belgian Bearded d'Anver Club
Belgian d'Uccle & Booted Club

White Leghorn bantam male as bred in Canada and the USA. Note the very large fanned tail and much smaller comb than on UK-type Leghorns. The Nederlandse Hoenderstandaard recognizes 'Nederlandse', 'Amerikaans' and 'Engels' type Leghorns as three separate breeds. The 'Nederlandse' type is midway between the others in terms of comb and tail size, not unlike 'utility-type' Leghorns in the UK and 'Italieners' in Germany.

Blue Andalusian & White-Faced Black Spanish Breeders Club

Chantecler Club of North America (probably has several Canadian members)

Cochins International (not known how 'international' it is)

Crested Fowl Fanciers Association

Cubalaya Club

Dominique Club of America

Dorking Club of North America

Faverolles Fanciers of America

International Cornish Breeders Association

International Heavy Duck Breeders Association

International Waterfowl Breeders Association ('international'?, may just include some members in Canada and Mexico)

Japanese Bantam Breeders Association

Modern Game Bantam Club of America

National Call Breeders of America (Call ducks)

National Frizzle Club of America

National Jersey Giant Club

National Leghorn Club

New Hampshire Breeders Club

North American Hamburg Society

North American Marans Club

Old English Game Club of America (for large OEG)

Old English Game Bantam Club of America

Oriental Game Breeders Association

Plymouth Rock Fanciers Club of America

Rhode Island Red Club of America

Rosecomb Bantam Federation

Russian Orloff Club of America

Sebright Club of America

Silkie Exhibitors and Breeders of America

Silver Wyandotte Club of America (Silver Laced, large and bantam)

Southern Cornish Club (US Cornish = UK Indian Game, club for south US states)

Standard Turkey Preservation Association

Sussex Club of North America

United Orpington Club of North America

World Cochin Society (not known how many members outside the USA)

Wyandotte Club of America

Wyandotte Bantam Club of America

Bibliography

Below are the books used to research *Exhibition Poultry Keeping*. Above these are some bibliographies that have been published for poultry-book collectors. The most sought-after, and therefore the most expensive books are the large volumes with many coloured plates illustrating the breeds, and specialized books that cover one breed only. Old books on diseases and commercial production methods are not yet (in 2004) generally collected. All are published in the UK unless otherwise stated.

Kuys, A., *Poultry Books*, Australia, 2002

Norris, John E., *Books on Poultry and Cock-Fighting*, USA, 1977

Palmer, John, *Books on Poultry*, Australia, 2003

Wright, Lucille N., *Books on Poultry Husbandry*, USA, 1961

Allcroft, W.M., *Incubation and Hatchery Practice* (MAFF Bulletin 148), 1964

American Bantam Association, *Bantam Standard,* USA, 1979; there are many more editions

American Bantam Association, *Book of Bantams,* USA, 1967

American Bantam Association, *Wyandotte Bantams,* USA, 1984

American Poultry Association, *The American Standard of Perfection,* USA, 1890, 1974, 1985

Anderson Brown, Dr A.F., *The Incubation Book*, 1979

Autosexing Breeds Association, *Autosexing Annuals,* 1946 to 1964

Baldwin, John P., *Modern and English Game Bantams*, USA, 1940, 1978

Banning-Vogelpoel, Anna C., *Japanese Bantams*, USA, 1983

Batty, Dr J., *Poultry Shows & Showing*, 1999

Bland, David, *Practical Poultry Keeping*, The Crowood Press, 1996

Bolton, W., *Poultry Nutrition*, MAFF Bulletin 174., 1967

Broomhead, William, W., *The Management of Chickens*, 1913

Broomhead, William, W., *Poultry Breeding and Management, circa* 1930

Brown, Sir Edward, *Races of Domestic Poultry*, 1906 (reprinted 1985)

Carefoot, Dr W.C., *Creative Poultry Breeding*, 1985

Caudill, David, *Araucana Poulterers' Handbook*, USA, 1975

Easom Smith, Harold, articles in *Poultry World* magazine 1968 to 1974

Easom Smith, Harold, *Bantams*, 1970

Easom, Smith, Harold, *Bantams for Everyone*, 1967

Easom Smith, Harold, *Managing Poultry for Exhibition*, 1974, 1980

Ellett, Arthur E., *Modern Wyandottes, how to Breed, Manage and Exhibit*, 1909

Entwisle, W.F., *Bantams*, 1894 (modern reprint editions also available, 1981 and others)

Fancy Fowl Magazine, Vol. 1, No.1, October 1981 to date

Feathered World yearbooks, 1910 to 1938

Hagedoorn, A.L. PhD, *Animal Breeding*, 1939, 1944, 1945

Hagedoorn, A.L. PhD, & Sykes, Geoffrey, *Poultry Breeding*, 1953

Hoenderstandaard, Nederlandse Bond van Hoender-, Dwerghoender-, Sier-, Wartervogel, 1986.

House, Charles Arthur, *Bantams and how to keep them*, circa 1930

House, Charles Arthur, *Leghorn Fowls, Exhibition and Utility*, 1927

Jeffrey, Fred P., *Bantam Breeding and Genetics*, 1977 (other editions entitled *Bantam Chickens*)

Jeffrey, F.P. & Richardson, W., *Old English Game Bantams...in the United States*, USA, 1991

Kay, Ian, *Stairway to Breeds*, 1997

Keeling, Julia, *The Spirit of Japanese Game*, 2003

Lamon, Harry M. & Slocum, Rob R., *The Mating and Breeding of Poultry*, USA, 1927

Mann, G.E., *Poultry Breeding*, MAFF Bulletin No.146, 1960

McGrew, T.F., *The Bantam Fowl*, USA 1899 (also reprinted edition 1991)

Nederlandse Wyandotteclub, *De Wyandotte*, various editions, Netherlands, 1998

North, Mack O., *Commercial Chicken Production Manual*, USA, 1978

Palin, John K., *Understanding Japanese Bantams*, 1980

Poultry Club of GB, *The Poultry Club Standards* 1886, 1901, 1905, 1910, 1922, 1923, 1926

Poultry Club of GB, *British Poultry Standards* 1954, 1960, 1971, 1982, 1997

Poultry Club of GB, yearbooks 1965 to 2003, especially the 1977 PCGB Centenary edition, the 2000 Millennium edition and the 2002 International edition

Poultry World magazine and annuals, 1907 to date (pure breed articles by Harold Easom Smith and David Hawksworth)

Powell-Owen, William, *How To Win Prizes With Poultry*, circa 1912

Proud, Pringle, *Bantams as a Hobby*, 4th edition 1912

Robinson, John H., *Popular Breeds of Domestic Poultry*, USA, 1924

Scott, George Ryley, *The Art of Faking Exhibition Poultry*, 1934 (also modern reprint edition)

Scott, George Ryley, *The Rhode Island Red*, 1939

Shakespeare, Joseph, *The Bantams Down-to-Date*, USA, 1925

Silk, W.H., *Bantams and Miniature Fowl*, 1951, 1974

Smith, Page & Daniel, Charles, *The Chicken Book*, USA, 1975

Somes, Ralph G., *Registry of Poultry Genetic Stocks in North America*, USA, circa 1975

Stromberg, Janet, *A Guide to Better Hatching*, USA, 1975

Stromberg, Loyl, *Exhibiting Poultry for Pleasure and Profit*, USA, 1978

Wandelt, Rüdiger & Wolters, Josef, *Handbuch der Hühnerrassen*, Germany, 1996

Wandelt, Rüdiger & Wolters, Josef, *Handbuch der Zwerghuhnrassen*, Germany, 1998

Weir, Harrison William, *The Poultry Book* (3 vols.), USA, 1905 (1st edn 1902, UK)

Williams, Rev. J.N., *Hamburghs in a Nutshell*, 1910

Wingfield, Rev. W.W. & Johnson, G., *The Poultry Book*, 1853.

Wiseman-Cunningham, R., *Gold and Silver Sebright Bantams*, UK, 1905 (reprinted USA 1977)

Wood-Gush, Dr D.G.M., *The Behaviour of the Domestic Fowl*, 1971

Wright, Lewis, *The Book of Poultry*, Popular Edition, 1888

Wright, Lewis, *The New Book of Poultry*, 1902 'Wright's Book of Poultry' revised by S.H. Lewer and leading specialists', 1919 (after Wright's death)

Wyandotte Club of Australia, *The Wyandotte Manual*, Australia, circa 1964

Index

Index